Transform Your Safety Communication

How to Craft Targeted and Inspiring Safety Messages for a Productive Workplace

By Marie-Claire Ross

TRANSFORM YOUR SAFETY COMMUNICATION

Copyright © 2013 by Marie-Claire Ross

All rights reserved worldwide.

Published by: Digicast Productions

Visit the author's website:

http://www.digicast.com.au

To receive a discount for multiple book orders, email the author: info@digicast.com.au

Advance Praise for Transform Your Safety Communication

"I recommend this Book/Guide to ALL Safety Personnel and Managers alike. It's very interesting and an easy to read guide that will encourage you to look at the way you communicate in a different light. Well done, Marie-Claire! **A thoroughly enjoyable read** and will now take the place of my dictionary as the most used book on my desk."

**Michael Carney, HSE Manager Sydney,
StarTrack**

"Simple sound theory backed up with experience, **filled with tips and examples** of the good, the bad, and the ugly of safety communication, finishing with a "how to" guide. For the busy person, the chapter summaries were perfect to get quickly to the practical implementation of Marie-Claire's marketing knowledge."

**Rachel Murphy, Health Safety and Compliance Coordinator,
IHBI Queensland University of Technology**

"It is not what you say, but how you say it." If you want to engage others and change their behaviour through effective communication then this book is for you. This book follows its own message in that; it is engaging, makes you think, and above all gives you a different perspective on how we, as people, see and hear what we want to see and hear. **I highly recommend the SELLSAFE system** as a toolkit that can assist you to improve safety and business performance."

**Paul Harper, CEO/Principal Mining Engineer,
AMC Consultants**

"Marie-Claire has distilled years of experience, effective marketing practices, wide research and wit into what **should become a safety professional's best friend**. These are techniques that work and applying them will bring your safety program to life – for me it's a communications adviser that's there when I need. This book is written in an easy to digest style with many examples and pictures to back up the content – the book is demonstration that the techniques work. I'm sold on safety!"

Alistair Camm, HSE Manager,
Pacific Aluminium

"I found the book **easy to read** and the safety information will be very useful to all working in safety. The information is written in such away that keeps the reader interested. I enjoyed reading this book and the conversation style, as it personalized the information."

Vicki Mutton, WHSE Coordinator,
SKILLED SA

"**Finally, a real communications book written for the safety professional**. Marie-Claire instructs us on a topic in a captivating manner that is essential for any safety professional who has struggled with communicating important safety messages to protect their workforce. I only **wish this book were written 15 years sooner**. I could have certainly used it to develop stronger safety cultures in the companies in which I have worked."

Morris Elkins, CSP, CPEA, CPSA,
Certified Lean Six Sigma Black Belt

"I recommend this work as offering something for every person who cares about effective safety communication. You will find at least one, if not many more, strategies to **transform your communication from ordinary to inspiring**."

Tamara Frigot, Regional SHEQ Manager,
Hyder Consulting.

"The tactics in this book are backed up by science, psychology, and tangible results. **Use the information to help you get buy-in** and agreement on the type of communication approach that is most effective in achieving positive behavioural change and keeping staff safe."

Alison Dillon,
Internal Communication Consultant

"**A powerful read**. If you ever thought you had safety communication and the challenges with it figured out, read this book, not once, but twice. The first time, the little bells will go off, the second time, you start to think about how much it all makes sense." Marie-Claire has done a wonderful job of writing about how to build a safety campaign so it is effective."

Rob Morphew, CRSP, EP Director, HSE,
Calgary Co-op

"Marie-Claire Ross's book, "Transform Your Safety Communication," is a fantastic resource for any safety professional passionate, in developing a successful safety campaign. **I wish I had come across this book earlier in my career as I found the simple marketing techniques**, examples, and templates invaluable for planning my next safety improvement project."

Christopher J Langley, National HSE Manager,
Higgins Coatings

"I really enjoyed reading "Transform Your Safety Communication" and was able to immediately implement the information into my safety management system. The information was well laid out and concisely written. I don't have any background in marketing and found the book to be **an easy read for someone brand new to the field**. The examples given were relevant and easily relatable. **This is a book that I have already recommended to my colleagues.** It's a valuable addition to my virtual bookshelf."

Carla MacKinnon, HSE Coordinator,
Pressure Services Inc

"Great leaders, coaches and managers all have the seemingly innate ability to create high performing teams by affecting their perspective and behaviour. As the adage goes - people buy what they believe and this book gives structure to that ideal. Understanding that you are selling an outcome does not belittle the cause and I don't believe safety is cheap, but the message has not been sold well. Marie-Claire is able to leverage off her marketing research experience on affecting human behaviour and serves up a process of applying this to health and safety. **A practical way to leading safety culture.**"

Mark Kerns, Managing Director,
Kern Health

To Andrew, Arielle and Amelie, my loving family, who teach me the power of communication every day!

Table of Contents

Chapter 6:
How to Make Your Safety Writing Engaging

PART VI: Templates

Chapter 7:
5 Templates to Make Safety Messages Stick

Preface

Fifteen years ago, I sat in the office of a rather exuberant public relations director of a public health department, as we carefully went through mock ups of magazine ads that had been prepared to communicate a new health website.

At the time, I worked in a market research company and we extensively tested a range of different ads that their advertising agency had created. We tested everything – the health content required for the website (from both doctors and the community), then later components of the ads – images, headlines and design. Our extensive research included focus groups, in-depth interviews, and telephone questionnaires.

This wasn't my first communication research assignment; for five years I had been testing countless products and services from cheese to shoe polish and even health insurance.

Fast forward to three years later and I'm running a video production house with my husband.

While sitting in the office of a production manager (also an exuberant person), I asked him how our training video was going for his production staff. What confused me was that he didn't use my pre and post questionnaires to test how well the training video went. It was just a video that he presumed worked well (and it must because it's still being used for training some 12 years later).

What surprised a market researcher let loose in the world was that people write information, create training videos, manuals, articles and - do NOT test them. Even more strange, was that best practices are never even considered (or shock, horror, known to even exist!).

Having worked with many passionate safety and risk professionals, I have always admired their tenacity to improve safety in their workplace. Yet, understood how frustrated they were that their communication efforts often seemed so futile.

Day after day, many of you go out amongst it all and try to transfer your enthusiasm to others. So gallantly laughing off accusations that "we've heard this before."

The reason I chose to write this book is that I want to help all the passionate professionals who realise how important it is to influence and engage on safety and risks. So many of you are keen to be a safety influencer, but have never been taught the required communication skills. Consider this **an easy go-to guide on the best practices to influence others on safety and quickly put together relevant communication**.

Too often, safety professionals are taught about compliance, but not the right skills to influence and engage others. This is so crucial when it comes to improving safety. After all, knowing rules and regulations is one thing, but getting people to follow them? That's something else entirely! Yet, so few companies even think about providing safety professionals with both verbal and written influence skills training to change outdated workplace behaviour.

Keeping workers safe is one of the most important things an organisation can do. I strongly believe that if you focus on safety, then all other things will fall into place, such as productivity and profits.

To make it easier for safety professionals to pull together attention-grabbing and memorable workplace safety communication, I have crammed this book full of lots of tips and templates. These are tested and tried techniques that work.

This provides you with easy shortcuts to quickly produce safety communication using all of the social psychologist techniques that advertising agencies use. Normally, you'd have to spend weeks searching through their information to get it to make sense to communicate on safety.

Instead, I've made it really easy for you to follow a proven formula to write interesting information on safety for:

- your company newsletter or intranet,
- training documents,
- senior management reports,
- safety meeting talks or presentations, and
- safety communication campaigns (e.g., safety themes).

This book is based on my years of communication experience in testing, creating and delivering marketing and safety training communication programs for clients and even, for my own company. Some of the programs I have worked on include manual handling training and awareness programs for both CSR Viridian and Gypsum Board Manufacturers of Australasia, truck driver training for Murray Goulburn and isolation procedures for a large steel manufacturer and Incitec Pivot to name a few.

I wish to thank those who kindly agreed to be interviewed for this book. Thanks goes to Chip Le Grand, chief editor of *The Australian* for his advice about creating leads and David Dumas from *Grey* for his input on the WorkSafe ad campaigns. Further thanks goes to Yarra Trams and the Helsinki Agency for permission to use "Beware the Rhino" posters.

A big thanks goes to our clients for agreeing to let us use some of the materials that we have created for them to show in this book. This includes GBMA, Rio Tinto (Pacific Aluminium), Pacific National and CSR Viridian.

I truly hope that this book will help companies around the world improve workplace safety communication, in order to humanise safety and ensure that everyone comes home from work safe.

Yours communicatively,
Marie-Claire Ross

Introduction

Effective communication is vital to get staff and contractors aligned and working towards a positive safety culture.

Yet, just providing training to work safely is not always enough. How **we communicate about safety influences whether or not people will accept or reject our safety messages.**

The main objective of any safety communication program is to change behaviour. **But how does a safety or risk professional change attitudes towards safety or improve the way people undertake procedures?** How can the safety manager deliver a message that motivates employees, supervisors, and contractors to think and act safely?

The secret to developing highly successful safety communication programs is to use marketing-based (also known as advertising-based) tactics.

But let's just pause and think about something for a moment. How do you feel about the term advertising?

With my blog, The Workplace Communicator, the only time I ever get negative comments is from safety professionals that believe that all marketing techniques are for con artists and that forcing anyone to buy anything is unethical. They often feel it's beneath them to have to use any marketers "tricks."

Of course, there is lots of shady advertising around. However, in this book, we're going to use the power of marketing for the greater good of safety.

First, let's get really clear on what advertising is all about. According to Wikipedia, *advertising is a form of communication intended to persuade an audience to purchase or take some action upon products, ideas, or services. Advertising can change attitudes, values and actions of those who see or hear the message.*

It's important to understand that the marketing techniques I want to go through with you are really **about influence to change behaviour, attitudes and values**. It's not about manipulating people and duping them into believing that they have to work safely (and here's a philosophical question for you: is it actually wrong to "manipulate" someone to work safely?) Instead, it's about persuading people to work safely, consider others, and learn new safety behaviour.

> "Advertising is only evil when it advertises evil things." David Ogilvy

What I love about safety communication is that at its core, it's all about saving lives. It's not forcing people to buy a product they don't need at a price they can't afford. Instead, when executed correctly, it's about getting people to understand why a particular safety message needs to matter to them and what they can do about it. **It's about positive behaviour change**.

In order to promote safety messages (and save lives!), it's time that safety professionals started to think like marketers.

So if you have any doubts about marketing, it's time to lose that queasy feeling. However, the fact that you have picked up this book means that you are keen to learn how to engage others on safety. Thank you for wanting to make a meaningful difference in your workplace!

If you are in a safety leadership role, this book gives you the tools to improve safety behaviour, reduce wastage and increase workplace efficiencies.

A healthy workplace culture emphasizes the importance of safety as part of how the organisation operates. When it comes to high performing businesses, there are three core inter-related areas that need to be in balance, in order to create the right environment for safety and productivity.

These are **Unity**, **Compassionate Leadership** and **Communication**. Once you get these three areas in balance, it produces trust (which is when people feel safe).

Workplace Culture Model for High Performing Companies

Communication
Transparent
Clear expectations
Predictable

Safety Ownership
High Candour

Unity
Group identity
Teamwork
Responsibility

Trust

Values
Emotional connections
Safety a Priority

Important contribution

Compassionate Leadership
Caring for staff
Clean workplace
Work/life balance
Safety integrated

Workers look to senior leaders to see that safety is a priority and that they are safe from harm. They get this from clear **communication** that is transparent, authentic and has no trace of hypocrisy. They see it from a **unified workplace** where people work together and look out for each other. And they feel it from **compassionate leaders** who care about them.

This in turn enables workers to feel safe to excel and take risks in their career, knowing they are protected from getting injured at work or from being bullied or harassed by colleagues (you can read more about improving your safety culture in this free report: www.digicast.com.au/workplace-safety-culture).

To be the best safety, executive manager, human resources, quality, risk or training professional, it's important for your career that you learn communication skills to influence, engage and improve workplace behaviour. The skills that you will learn from this book are critical for you to improve safety in your company and achieve your goals. They can also be transferred to other areas in your life.

With that in mind, let's take all of the ethical parts of advertising and use them to create a safe, happy workplace.

To your communication success!

How to Use This Book

I know you're all busy and writing a safety communication campaign is probably your least preferred task in the world.

So to make this learning process foolproof, I've written this book so you can easily scan the content and find relevant information. You do not have to read every page (although, it's better if you do).

In general, I make reference to safety professionals in this book.

However, I know a lot of you are not safety professionals all the time; some of you are in administration, senior management, human resources, internal communications, quality and environmental departments (sorry if you're not in this list, you can always email me and tell me what you do at info@digicast.com.au).

However, the fact that you are reading this book means that you are not like the average safety professional.

Truly great safety professionals don't blindly follow instructions, they challenge the status quo and look at ways to better communicate their message. They're not happy with mediocrity. They live for their job and are always thinking of innovative solutions to their safety problems. They don't just accept the standard answers. At any given opportunity, they consider options to ensure workers make it home to their families.

And this is called art.

Let's get things straight. Art doesn't mean you can draw or paint, it means you can see. You can see what's right or wrong.

As Seth Godin makes clear in the wonderful book, "Linchpin", art is anything that's creative, passionate and personal. **And great art resonates with the viewer, not only the creator.**

Artists can work with watercolour and clay. But artists can also work with business strategies, customer service, managing a meeting and safety. **Art is about intent and communication, not substance.**

Artists take their job personally. They are willing to challenge the status quo, be bold, insightful and creative.

A safety artist is not willing to put up with ordinary safety results. He (or she) is ready to do what it takes to inspire a workforce to work safely. He might have to be compliant, for compliance sake, but he's also prepared to do more than just be compliant.

> *"Art is a personal gift that changes the recipient. The medium doesn't matter. The intent does."* Seth Godin

In the book, "Grow," Jim Stengel believes that great CEOs are actually brand artists that connect people holistically with both rational and emotional information. They lead people with unified stories that communicate how the company improves lives both internally and externally.

What I want you to understand is that no matter whether you're the CEO or Safety Director, to get the best results, you need to consider yourself as an artist.

This book will transform your safety communication so that you start to operate on an entirely different level. No longer will safety be seen as grumpy talk from an overbearing policeman. Instead, this book will teach you how to align everyone on the importance of safety within the company.

To make it easier, as this book is about safety communication, I'll refer to you all colloquially as safety professionals or safety artists; as when you are writing safety communication, that's technically your job role.

The objective of this guide is to make it really easy for safety professionals who are new to the topic of safety communication, as well as those who have been communicating safety for some time and want to learn new techniques.

This introductory content includes step-by-step instructions on how to get started with some communication methods to change behaviour, as well as lots of easy to use information and templates to start crafting your messages.

After reading it, you will be able to execute basic and medium-level marketing tactics to improve how you engage others on safety.

How this Book is Written

Essentially, this book is broken down into six sections:

 1. **Learn**: Discover the Proven Safety Communication Framework

 2. **Attention**: Capture Your Audience's Attention

 3. **Remember**: Achieve Memorable Safety Communication

 4. **Persuade**: Get the Effective Communication System

 5. **Create**: Transform your Safety Communication

 6. **Templates**: 5 Templates to Make Safety Messages Stick

Chapters 1-4 provide you with theory and examples of how to create compelling safety communication that will grab attention, be memorable and positively influence people to change behaviour. At the end of chapter four, I introduce the SELLSAFE formula for effective safety communication and delve into some examples and how Grey, the ad agency, creates their award-winning safety campaigns.

While chapters 5-7 are all about how to create your safety communication with lots of tips and templates. All of the theory that you have been taught is combined into easy to use models. This is the action chapter and it's where you can start "doing." It is where you can get up and running quickly.

There are two types of safety communication that a safety professional creates:

1. **Short Communication Pieces** – For many of you, there will be times when all you need to do is write a short safety communication piece such as an article for a newsletter, a report for your senior manager or workers, or a speech/presentation for a meeting. If you are short of time, skip straight to Chapter 5, read the tips and follow the Safety Communication Content Creation Template.

2. **Long Communication Pieces** – This is when you have to create a safety theme or introduce a new safety training process that will be repeated over time (usually 1-3 months). **This is what is referred to as a workplace safety campaign.** This type of communication is like a full ad campaign with a slogan, images, lots of different communication mediums (i.e., emails, posters, training), as well as a schedule of communication events. Ideally, read the whole book, in particular the first four chapters on how to change behaviour. The three ad templates at the back of the book are the most relevant to you, as well as the Communication Schedule.

Throughout this book, you will find "Sticky Tips" to help reinforce important information to make your messages stick.

Now, safety artist, it's time to create a world of safe workplaces, let's get started.

PART I

LEARN

Chapter 1: Discover the Proven Safety Communication Framework

Let's think about television commercials. Despite the fall in television watching in recent years, and the subsequent drop in advertising revenue, advertising still has a proven record of changing consumer habits and beliefs; all while helping countless companies sell products.

But it's not just products that advertisers have successfully convinced us that we need in our lives. Australian Government organisations like WorkSafe and VicRoads have used television commercials successfully in the past to change our behaviour and attitudes towards workplace safety and road safety respectively.

Back in 1989 when VicRoads introduced the "Drink Drive, Bloody Idiot" advertising campaign, it encouraged people to start thinking about drink driving and to even become against it. Before then, many people's attitudes towards drink drinking were that it was okay. Now, Victorians believe driving a car while over the alcohol limit is totally unacceptable.

While in the United States, the Ad Council also created a series of public safety campaigns that have changed behaviour and subsequently, saved thousands of lives. Safety Belt Campaign is one of the most famous for having increased safety belt usage from 14%

(in 1985) to 79% (present), saving an estimated 85,000 lives, and $3.2 billion in costs to society.

So do like the experts do and use marketing techniques to sell your safety messages. Advertising informs and reinforces the need for safe practices. But advertisers know that you just can't say your product is the best. Likewise with safety: you can't just say your company safety message is important and leave it at that.

There are four key building blocks to a safety campaign that I want you to use for all of your safety campaigns (both large and small). Consider these to be the **Four Commandments of Safety Campaigns**. They are:

3. Promote your message in **multiple places, multiple times.**
4. **Define your target audience.**
5. **Communicate one clear message.**
6. **Maintain a consistent look and feel.**

Let's go into these in more detail:

1. Multiple Places, Multiple Times

Throughout the course of a day, people are constantly bombarded with marketing messages. Estimates vary from around 150 - 5,000 messages every 24 hours.

Successful ad campaigns have to compete with many other goods and services to grab the attention of people. In advertising speak, it's important to "cut through the clutter."

Back when I used to monitor the success of advertising campaigns to report back to ad agencies as to how well they were performing, one of the questions we used to ask people was how many times they had watched a certain ad. Known as *frequency*, advertisers recognise that people need to be exposed to a television ad 4-7 times before they will absorb the message. **This is why frequency of message equals success in advertising.**

It is such an important measurement because the more times a person has seen an ad, in whatever format (be that an email newsletter, poster, radio or television), the more likely they will be to recall it. And be convinced by the subject matter.

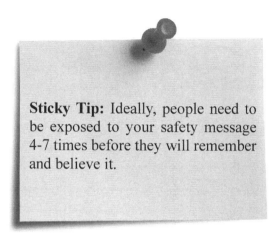

Sticky Tip: Ideally, people need to be exposed to your safety message 4-7 times before they will remember and believe it.

As a safety professional, your communication messages compete with messages from the production manager pushing for better productivity, the human resources manager needing forms filled out more accurately, and co-workers distracting each other. And then there are messages from home that you have to battle with such as family issues, social networking sites, money problems and of course, other advertising.

That's a lot of additional messages that you have to compete against! To ensure that your workers remember your safety message, you need to get into their head – a lot. This means planning on **multiple message placement.**

For example, if you have a monthly safety theme, you must repeat your message in multiple places and formats.

Plan to engage workers with your safety message in a variety of ways, keeping in mind that people learn by watching (visual), hearing (audio) and touching (e.g., getting people to put on sunscreen for a safety theme on sun protection). Different communication methods include.

Videos	Photos	Posters
Email Newsletters		Screensavers
SMS	Checklists/ Handbooks	Toolbox Talks

When putting together your safety campaign you will need to plan a schedule of daily, weekly and monthly messages to staff. For example:

activity	supporting collateral required	Aug 1	2	3	4	Sep 1	2	3	4	Oct 1	2	3	4	Nov 1	2	3	4	Dec 1	2	3
Initial presentation	PPT slides with video content	■																		
Visual reminders	Posters			■																
Personal copy to all at second presentation	"Fold-up" pocket card					■														
Reinforcing letter sent home from GM	Letter							■												
Email newsletter	Story about staff member undertaking the new initiatve correctly									■										
Staff lunch time event	Demonstration of correct safety process											■								
Reinforce risks and rules	> Stickers to place on / near bins													■						
Publish workgroup success stories	> Communicate successes															■				
Reinforcing letter from GM & pocket calender sent home	> 2011 Pocket calender																	■		

Most safety training programs fall short when it comes to frequency of message. It is like glue that holds the tactics together and is essential in successful advertising. All it takes is planning.

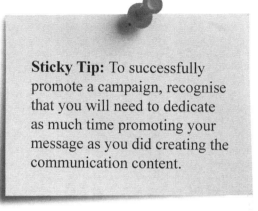

Sticky Tip: To successfully promote a campaign, recognise that you will need to dedicate as much time promoting your message as you did creating the communication content.

2. Define Your Target Audience

Another important advertising strategy is to customise your communication to your audience. This will ensure that you create relevant safety messages to all of your different workplace groups

While we will go into this in more detail later in the book, it's important to define your target audience.

For example: *male workers aged 25 – 55 years.*

Based on this, you can work out the communication formats that best work for them. It is recommended that you test this out to discover which form of communication is the most successful at getting through to workers. This can even include testing such as determining which locations work best for posters.

For example, my teenage employees prefer SMS messages to email (and some really tricky previous employees only seemed to respond on Facebook). While my mature staff always respond to emails and ignore texts! We have two main forms of written staff communication: SMS and email.

For medium to large sized organisations, you will need to tailor your communication to the formats more suitable for your audience.

3. Communicate one clear message

Again, this is another item that we discuss in more detail later in the book, but with any safety campaign, it's crucial to only communicate one main message at a time, and to consistently promote the same standardised safety message.

When we communicate, we often try to tell everything right up front with perfect accuracy. However, this makes it hard for people to comprehend, as we can only grasp the meaning of one message at a given time. Give people too much information at once and confusion sets in, resulting in them tuning out pretty quickly.

The solution is to give just enough information to be useful, in order to grab attention. Then you can provide a little more and then, a little more.

Having said that, it is really difficult to focus on one core message. Even experienced marketers end up throwing too many messages into the mix.

Over the years, I've worked with lots of marketers to create marketing videos for their new products and services. Often, because they are so close to the information, they will provide me with around 7-10 messages.

Obviously, you can't create a short video with that many benefits, so I would work with them to whittle it all down to their core message. You'll be amazed to know that is a very hard job. However, as an outsider it was easier for me to distinguish the core than them (provided I wasn't overly submerged in the information).

By focusing on one core message, it also makes it hard for people to misinterpret what you mean or walk away with a different meaning. It ensures you create fresh and clear communication.

We will go into more detail about how to focus on one message in the chapter **Remember**. For now, I just want you to absorb that you need to focus on creating one main safety message.

4. Consistent Look and Feel

It's also really important to have the same look and feel throughout your marketing campaign.

By using the same colours, font style, language, message and design, people will start to instantly remember the meaning. This is because it creates a mental model for readers and enhances their understanding of the topic. It is actually known as a graphic identity.

For example, those of you who live in Melbourne, Australia (or have visited) will be aware of what this means:

This is a part of a safety campaign that Yarra Trams (created by the Helsinki Agency) in Melbourne used to warn pedestrians about just how heavy trams are, in order to stop people from walking in front of them.

The main part of the campaign began with this ad (http://www. digicast.com.au/beware-the-rhino/) that was combined with posters that were plastered all over tram terminals.

What is interesting is that the colours, style and message are always consistent. This safety campaign has proven to be highly effective and received a high 84% of prompted awareness (back when I was a market researcher, most of the ads we measured averaged around 30% awareness; a result this high was always cause for amazement).

Worldwide, successful safety campaigns include a variety of elements (outdoor, radio ads, digital, TV ads and social media) with the main message being clear and displayed consistently.

For your workplace safety communication, make sure that all of your safety campaigns or themes look similar. Never let a department modify any portion of your safety communication to change the colours or messages. Remember, "Consistent Look and Feel" is a commandment and must be kept sacrosanct.

Incidentally, this also means ensuring that all of your different departments or sites are aligned with the message. You do not want them to send out conflicting information (e.g., safety officer tells people to work safely and cautiously, but production manager pushes for speed).

By always using these four advertising commandments, you will start to grab the attention of people with your communication and help them to understand your message. Of course, it's a little more complicated than that. The next few chapters go through in more detail how you do this.

Summary Tips

Learn

- Plan your safety campaign to include a range of formats and methods. You can't promote your message enough! More promotion is always best.

- Tailor your communication to your audience. Test what works best for different demographics (i.e., do workers aged 18 – 25 years respond more to SMSs than emails?)

- Create one clear safety message

- Use the same consistent look and feel

PART II

ATTENTION

Chapter 2: Capture Your Audience's Attention

Why does Safety Communication Get Ignored?

You know what it's like; on a wall in your main corridor at your workplace, you've got a poster that's going slightly yellow with a safety message that people ignore.

Or even worse, is that you tour a workplace site and find workers ignoring safety equipment and devices.

Despite your best intentions to communicate new safety themes each month to your employees, they seem to barely remember last month's safety theme.

Essentially, throughout our day we are bombarded with lots of stimulus – dogs barking, neighbours yelling, phones ringing, air conditioners humming, conversations all around us and of course, advertising messages fighting for our attention.

Our brain has learned to adapt to our environment by filtering out what we don't need to know about, to avoid information overload. As a result, our brains have evolved to selectively screen out noise and visuals.

Say you have decided that you want a new car and it has to be red. You'll find that every time you go out, you will literally see red cars everywhere and be surprised at how popular they are (the same goes if you've decided on a particular model of car). This is because in the past, you filtered out red cars because your brain had other things to focus on. Now, that you've told your brain that you're interested in red cars, the filter is removed and the invisible becomes visible. You see red cars everywhere!

Psychologists dub this the 'The Scotoma Effect' or RAS (Reticular Activating System). The word 'scotoma' means 'a mental blind spot, or a gap in the field of vision'. This means that if you are not looking for red cars, your brain will delete it from your reality by creating a scotoma, which means you're blind to red cars.

When you're writing a safety campaign, you are writing it for scores of workers who at that time are not particularly interested in ladder safety. In fact, they've deleted it from their brain because they're actually more interested in working out how to get their teenager to stop skipping school.

Memory, interest and awareness influences what gets our attention.

In the book, "Brain Rules" by John Medina, he claims that our brains constantly scan the sensory horizon, continually processing whether or not events are important enough to notice.

Every day, we use previous experience (memory) to drive awareness of what's important or not in our environment. We create patterns or constructs from our everyday life to help us quickly adapt to what is around us. We have to, so we don't get overwhelmed.

So how do you get around this clever brain filtering to get people's attention?

As Chip and Dan Heath say in "Made to Stick," our brains are designed to be aware of changes. One suggested theory is that when our ancestors used to roam the wilds of the African savannah, they had to constantly assess the environment for possible danger and to ensure they were safe. To this day, we are finely tuned to assess potential threats.

We only notice things when something changes such as the air conditioner turns off or a house alarm starts screeching.

That's why advertising that features novel, unfamiliar and unpredictable stimuli grabs our attention. You've got to break a pattern.

Emotions get our attention. Any event that heightens our emotions is much better remembered than a neutral activity.

In particular, there are three emotions that you need to leverage in your safety communication to make your safety messages sticky.

Give people a *surprise*, induce *fear* or provide *interest* through fascinating and unusual information.

Surprise Your Audience

Take a look at this television ad: http://www.digicast.com.au/book-resources/

Warning – Spoiler Alert! Please watch the ad. If you really are unable to watch it, then this is what happens:

It starts with a family walking towards their Enclave minivan. They are then in the van, cheerfully playing with all of the cup holders and the sunroof, while the narrator breezily lists all the amazing features ("It's got temperature-controlled cup holders"). After seeing different angles of the vehicle, the camera angle shows the reflection of the ten year old son, smiling through the window.

At that moment, we hear tyres screech and then we see a car hit the van side on, in a terrible collision. The ad fades to black and the title appears "Didn't see that coming?" The question fades and is replaced with "No-one ever does," followed by "Buckle up. Always," with the sound of the jammed horn blaring in the background.

According to Chip and Dan Heath, **why this ad is so effective is that it gives our brains a big time surprise**. What appears to be a normal car ad, is in fact a strong message to wear safety belts.

It caused our guessing machine to fail.

Psychologists label prerecorded information that's stored in our memories as "schemas." If someone tells you that they've just bought a fantastic leather jacket, an image might spring to your mind with something like a black, biker's jacket *á la Fonzie* from Happy Days.

With the Enclave ad, our schema predicted that it was a routine family car advertisement. However, our concept of how the world works was challenged, resulting in us being surprised. **When our guessing machine fails, surprise grabs our attention, so that we can learn for the future.**

You could think of surprise as a momentary emergency override that renders us speechless while we try to process what happened.

When it comes to creating memorable safety campaigns, it's the unexpected ideas that create surprise, that stay in our memory. And grab our attention. This is when people start to consider changing their behaviour.

What's so handy with generating surprise is that it prompts us to hunt for underlying triggers, to imagine other potentials, so that we can avoid such a surprise again. It's this process that helps us to remember information.

Interestingly, when we go into surprise (or shock), we become highly suggestible. We are more likely to follow a direction after being surprised.

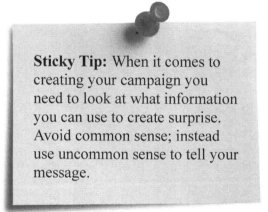

Sticky Tip: When it comes to creating your campaign you need to look at what information you can use to create surprise. Avoid common sense; instead use uncommon sense to tell your message.

Ads that get your Attention

I've put together a selection of "surprise" ads for you to watch. I won't tell you anything about them, so as to not ruin the endings.

http://www.digicast.com.au/surprise/

The Power of Fear

Fear in communication can be highly effective, but only if you give specific steps so that people know how to avoid the threat. People need to believe the action will adequately remove the danger and that they can easily perform the action required.

If you just scare people without advising on how to sidestep the risk, fear appeals can backfire, which can even lead to a denial of its existence. This results in inertia that makes people literally too scared to take any action.

As an example, watch this absolutely atrocious safety video: Will You be Here Tomorrow? http://www.digicast.com.au/blog/ bid/85671/When-Workplace-Safety-Training-Videos-Go-Bad

The video features enough gore and bloody industrial accidents to make anyone seriously question why you would even want to go to work. It's really off-putting and difficult to take seriously.

While safety has a strong history of using over-the-top fear, the top category would have to be political ads, which have been instilling terror for 50 years or more.

Take a look at two political ads (link below). These classics were both extremely successful at helping their candidate win by a strong margin. Notice how they advised voters on what steps to take by ending with clear voting instructions.

http://www.digicast.com.au/political_ads/

However, when it comes to fear, what humans are really trying to do is **minimise loss aversion.**

Research shows that people actually dislike losing more than they like winning. Therefore, losses loom larger than gains even though the value in monetary or other terms may be identical.

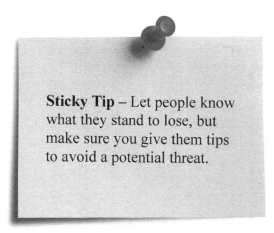

Sticky Tip – Let people know what they stand to lose, but make sure you give them tips to avoid a potential threat.

Keep Things Interesting

New, interesting information that we haven't heard before often gets our attention.

"We don't pay attention to boring things."
– John Medina

Gossip is a prime example that has created a thriving gossip magazine culture. It's even believed that we're hard wired to gossip, as in the past we needed to know about everyone in the village to ensure reproductive success.

However, while gossip is an attention-grabber, it's only useful in safety when you use it to talk about how other people or companies are dealing with safety.

In the next chapter, **remember**, we will go through a range of steps on how to make your safety content more interesting. For now, all you need to know is that providing interesting content is a key requirement for your safety communication.

The Need for People to Like You

As Robert Cialdini references in his book, "Influence: The Psychology of Persuasion", to be able to influence and get people to listen, they actually need to like you.

This could be a delicate issue for some safety professionals. After all, there is a common perception that safety artists are a bit like grumpy policemen. Always telling people off and never saying anything nice.

While I'm sure a lot of you feel the burden of having to be 'bad cop', it will undermine your ability to change behaviour.

What Cialdini found is that we tend to like people who:

1.   Appear similar to us.
2.   Give us compliments.
3.   Are working with us on the same goals.

It is much easier for employees to buy into your message if they like you. This means getting out on the floor and talking to people or having a break with everyone in the tea room. People need to believe you're at their level, before they will listen to you.

Start taking an interest in your workmates. Ask questions about their family and their hobbies. This will make your job easier, as people will work with you, not against you, if they know you're on their side.

There will always be times when you won't be able to get everyone to like you. In these circumstances, it's more important that you are well-respected and workers know that you are sincere in your desire for a safe workplace. **They know they can trust you.**

As Peter Drucker said: "trust is the conviction that the leader means what he says…a leader's actions and a leader's professed beliefs must be congruent, or at least compatible."

As I mentioned previously in the "Workplace Culture Model for High Performing Companies", once you balance clear communication with compassionate leadership and unity, you hit the sweet spot. Trust. This is so important for a happy and productive workplace.

If you feel that you are in need of help, we have created a training program for supervisors, managers and safety professionals. The program teaches key influence and leadership skills to improve safety that will ensure you lead a happy team (that will buy your message and do what your request because they like you).

Visit: http://www.digicast.com.au/toolbox-training-DVD

Summary Tips

Attention

1. The brain is constantly bombarded with information to ensure that it doesn't get overloaded; it develops shortcuts for quick decision-making.

2. Our brains are designed to be aware of changes. We only notice things when something varies such as the air conditioner turns off or a car alarm starts shrieking.

3. **Advertising that features novel, unusual and unpredictable stimuli grab our attention. You've got to break a pattern.**

4. Unexpected ideas are more likely to be remembered because surprise makes us pay attention and think. It "sears" the unexpected onto our brains. Avoid being predictable.

5. If you want your safety message to be memorable, you've got to break down people's guessing machines and then fix it. Disrupt people's constructs about the world. Find out what's counter-intuitive about the message and get people thinking about it.

6. There are three emotions to leverage in your safety communication to make your safety messages sticky. Give people a *surprise*, induce *fear* or provide *interest* through interesting and unusual information (think gossip and conspiracy theories).

7. **Common sense is the archrival to sticky safety messages.**

PART III

REMEMBER

Chapter 3: Achieve Memorable Safety Communication

What do you *Meme*?

Can you recall when you were at school and you were taught little sayings to help you remember information?

For example:

> I before E, except after C.
>
> ***Emma*** faced a **dilemma**.
>
> There's a ***rat*** in sepa***rat***e.

These handy little grammar and spelling rules are what are called **mnemonics.** Essentially, this is a device, such as a formula or rhyme, used as an aid in remembering.

While they work well for children, they work even better for adults.

In the article, "Stalking the Wild Mnemos: Research That's Easy to Remember," educational psychologist Joel R. Levin found that "sufficient research evidence now exists to suggest that even skilled learners can become more skilled through mnemonic strategy acquisition and implementation" (1996).

One of the things safety professionals tend to like are safety slogans. Whenever I write a blog article about catchy safety slogans, I know that I will get a lot of positive comments and a few new ones to add to my list.

What safety professionals get on an intuitive level is that safety slogans are an important way to break through all of the clutter (but only good safety slogans, not lame, boring ones).

In the book "Rapid Response Advertising," by Geoff Ayling, he also discusses a biological theory that when humans are under pressure, they are unable to think clearly, so the brain relies on mental shortcuts that can save time (and possibly life). Humans rely on a range of different mental shortcuts to make life easier, that we are often not aware of.

Mnemonics work so well because they are mental cues in all kinds of learning situations. Essentially, that's what a great safety slogan is – an easy way to memorise how to do something.

In advertising, mnemonics are often referred to as *memes*.

Ayling says that "A meme operates through the process of chunking complex concepts or ideas down into simple, easily units".

You see advertising memes everywhere. In fact, you've probably got a whole lot stored in your memory that you didn't even try to memorise.

However, memes themselves are a little bit complicated and often misunderstood. This is because a lot of people are unclear on the definition of an actual meme, which is to tell people what the product offers or how to do things.

Advertising Meme Examples

"When you absolutely, positively have to have it overnight."
(Federal Express)

"You're not you when you're hungry." (Snickers)

"M&M's melt in your mouth, not in your hand."

"Have a break, have a Kit Kat."

They're not to be confused with aspirational brand straplines or taglines which do not describe what the product will actually do for you.

Just do it. (Nike)

Think Different. (Apple)

Because You're Worth It. (L'Oreal)

Marketing guru Robert Middleton says that memes accomplish four things. They:

1. Actively transfer specific information.
2. Are immediately and obviously beneficial.
3. Are self-explanatory and ultra-simple.
4. Are easy to replicate in someone else's mind.

Using memes in your safety campaigns helps people understand and remember information. Great safety memes strive for clarity, authenticity, and simplicity. They provide a memorable and direct connection to information.

Great safety slogans must:

1. **Be positive** - Avoid creating a slogan that focuses on behaviour that you don't want. Instead, write a safety message that conveys what you want people to do. For example a negative slogan for height safety is "Don't fall for it". Using more positive language, a more appropriate version is "**A harness is better than a hearse**". While this might have negative connotations, it still focuses on what you want the person to do, rather than the wrong behaviour.

2. **Keep it short (and tweet)** - In this age of Twitter, being

able to write in 140 characters or less helps you to condense your message. It's the same with writing a safety message; just try and encapsulate it in eight words or less. Avoid long and complicated safety messages like "Don't Fall Asleep At Work and Get Your Head Caught In a Splicing Bar" (too long, negative and not even funny) and "Mine Eyes Have Seen The Gory Of The Coming Of The Blood, It Is Pouring Down My Forearm In A Bright Red Crimson Flood" (ditto).

3. **Avoid jargon** - Make sure the sentence flows easily. Avoid acronyms and words that not everyone will understand. To check comprehension, ask yourself: "will my mum get this one?"

4. **Contain a surprise** – As mentioned, previously **common sense is the adversary to sticky safety messages**. When our brain's guessing machine fails, it wants to work out why it was unable to guess. This surprise grabs our attention, so that we can be prepared in the future. By trying to work out what went wrong, our brain is more likely to remember the information.

Here's a good example (a personal fave):

Hug your kids at home, but belt them in the car

Slogans that contain the obvious will be ignored. Examples are: "Play it safe" and "Be aware, take care". Yawn! Avoid being predictable.

5. **Play on words** - A clever play on words helps to make your safety message just that little bit more memorable. This can include rhyming and repeating words in a different order. Adding a little bit of fun can make a serious subject more approachable. Personally, the funnier the better (but I love Dad jokes, so the funny bar is pretty low for me).

6. **Contrast** - Provide a contrast such as a do and what not to do. For example:

Prepare and prevent instead of repair and repent

Great safety memes (slogans) that give information on what you want people to do in a memorable way include:

Lifting's a breeze when you bend at the knees.

While on a ladder, never step back to admire your work.

Housekeeping you skip may cause a fall or slip.

Sticky Tip: Only use safety slogans (memes) that tell people what you want to do in a memorable way. Avoid being predictable.

Using a slogan effectively is a bit tricky. That's mainly because a lot of them are not true **mnemonics,** which are the **key to learning and storing information in memory.** By using a true meme, you'll be able to use safety slogans to your advantage.

Ideally, use slogans at the end of your writing or speech, so that you create an **impact line** that cements the information.

Get to the Heart of the Message

After Michelangelo had created David, he was asked how he did it. He answered: *"It's simple. I just removed everything that doesn't look like David."*

When it comes to safety communication, you have to drill down to the important information that grabs attention. Throw out everything that is meaningless and unnecessary.

This means you have to master the art of exclusion and strip down your safety message to the essential. You need to get to the heart.

From a psychological perspective, the more information you give people, the harder it is for them to make a decision on what to do and how to prioritise what's important.

It also means the more likely people will misunderstand what you are trying to communicate.

As Chip and Dan Heath say in "Made to Stick," by promoting and repeating one core message, it helps people avoid making bad choices by reminding them of what's important. **The more we cut down the information into a single idea, the stickier it will be**.

Ideally, you need to condense your safety message into one core sentence, but short is not the aim. Rather, **the objective is to make it profound and simple to understand.** You don't want to dumb it down too much.

This isn't as easy as it sounds, because as humans we have a few flaws that stop us from communicating simply.

Most people like to over-complicate things, without even realizing it. Being able to make the complex, simple, is a skill.

The key to effective written communication is that people must be able to understand it.

> *"Write to the chimpanzee brain. Simply. Directly."*
> - Eugene Schwartz

Often, when we have trouble working out the "heart" of what we want to say, it's because we're not clear in our head. This either means we have more work to do or that we need to get someone else involved to work it out for us.

Burying the Lead

When it comes to writing stories clearly and succinctly, newspaper and magazine journalists are pros. After all, they know too well how quickly they need to grab the attention of readers.

In a news story, the "lead" is the opening sentence. Often, it is written in a "who", "what", "where", "why" and "when" (the five W's) format to enable the reader to quickly scan all of the information they require and decide whether they need to read on.

Writers are often admonished "Don't Bury the Lead!" to make sure they present the most important and interesting facts first, rather than requiring the reader to have to search through several paragraphs. Or even worse, try to work out what the main story is, in the article. This is considered a big mistake in journalism because if the lead isn't clear, people will stop reading.

A good lead will attract the reader and motivate them to continue. It encourages the reader to ask themselves more questions, so that they want read to on to find out more.

For the writer, the tricky part is actually distinguishing the lead, as there are so many facts to sift through in every story.

Chip Le Grand is the chief reporter at "The Australian." With 20 years of journalism under his belt, he says that many journalists will take as long as it takes to get the lead right. In fact, journalists will spend most of their writing time fine-tuning their first sentence (or lead).

In Chip's words: "When working out your lead, there are two things you need to do. You need to first work out what you want to say, then, how to say it. What's the most interesting thing? How do you do that succinctly? You keep revising to get rid of clutter. Essentially, you want your lead to be read in one breath.

There are two types of leads. The first one is where you describe a scene in word pictures. Think of a WorkSafe safety ad. It always shows everyday scenes people can relate to. You want readers to recognize themselves in it. The classic news style is the 5 W's, but it's more important that you write in simple sentence constructions with few commas. In fact, we focus more on 4 W's now, as the when is not so important."

For example:

James Hird has agreed to do the time but not pleaded guilty to the crime.

The interesting thing with this lead is that it makes you think twice about your assumptions because it's unusual for someone to agree to do time, but not actually admit guilt.

Another short, but more visual example is:

In the early hours of this morning, homicide detectives were led to a dark field on the edge of Melbourne, to the body of Jill Meagher.

This style of lead is more story-based and prompts you to ask other questions like: what suburb? How did she get there and who did it?

While this longer lead is more of a traditional format:

James Hird was last night considering whether to step down as Essendon coach and clear the way for the club and the AFL to resolve the supplements scandal that has engulfed the season.

Essentially, you want to tease the reader with enough contextual information that encourages them to ask more questions.

There will be times when you do not need to use all of the 5W's. A condensed version is called the "What, Why, How" formula:

1. **What** – state what the issue is about.
2. **Why** – explain why the issue is important to know.
3. **How** – provide information on how to apply the new information.

Sticky Tip: Always start your safety communication by asking what's the interesting information? Use your core message, the 5W's or the "What, Why, How" formula to encourage people to want to find out more.

Inverted Pyramid

After the lead, journalists write the information in what they call an "inverted pyramid."

The most essential and interesting pieces of information are at the top, while supporting information follows in order of importance.

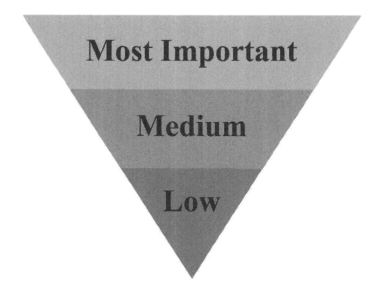

When writing safety communication, you need to include the most essential information in your first 1-2 sentences and then use the supporting information in subsequent paragraphs.

In "Brain Rules," John Medina discusses that when our emotions are awakened we remember the gist of the experience, but not the details.

Our brains process meaning, before detail. **Therefore, by providing your core concept first, you help people understand information by a high 40%.** The brain likes hierarchy. Start with the summary information (or your lead – who, what, where, why, when) and then explain additional information in a hierarchical manner.

Make sure you include surprising facts and avoid using gimmicks (e.g., a surprise for the sake of it with no real connection to the message).

The Curse of Knowledge

You can't unlearn what you already know. Once we know something, we can't imagine not knowing it. Our knowledge has "cursed" us, as we find it difficult to share it with others, as we can't readily explain what we know.

According to psychologists, the **curse of knowledge** is a cognitive bias, which makes it hard for experts on certain topics to be able to think about problems from the perspective of the lesser-informed person.

For example, take teaching your child to drive. I remember as a 17-year-old receiving driving lessons from my Dad.

My father was teaching me how to drive a manual car and exclaimed that it was really hard for him because he no longer drove thinking through all of the steps. Having to retrace all of the stages required to drive a car was challenging. I remember just having to sit and wait beside him in the car while he mumbled to himself as he went through the steps backward.

In business, leaders have big visions and ideals of where they want the business to go, yet have lots of trouble extracting their ideas into information people can understand, in their business strategies and visions. Too often, CEOs focus on information such as "maximise gross profit" without making that meaningful to employees.

Later on in the book, we will go into an eight-step formula on how to get around "curse of knowledge" bias. But for now, just realise being an expert on a subject actually makes it harder for you to communicate to those who know nothing about the topic.

Using Comparisons

Geniuses like to go into lots of detail. This often confuses and overwhelms their audience with people reacting by disagreeing to the information.

To help people understand new information, it's often beneficial to compare the new information to something they already know about.

Instead of sharing unknown information, with other unknown information, you need to connect the unknown with the known. Otherwise, you'll switch off your audience.

For example, if I told you that a mandalo is a fruit that tastes a lot like mandarin, you'll know immediately that it would be from the citrus family and that it would probably work well in a cocktail.

Again, by understanding that humans tend to learn by organising information into schemas, or patterns, you can leverage this to build new information into their constructs of how the world works.

It's important to create a reference to something your audience already associates with. This provides a feeling of familiarity that helps understanding. It also stops people from complicating things, resulting in them ignoring what you say.

When trying to shorten your core message, compare your content to something easily familiar. Then, you can sequentially add more information.

As John Medina says in "Brain Rules," to help people recall information, we need **to provide real world examples, as it makes the information more complex and better encoded.**

One of the areas that people have difficulty comprehending is dietary information on specific recommended daily allowances. This is probably because it is difficult to get clear on what a healthy diet even looks like.

For overweight people, it's incredibly hard for them to understand how much they are overeating, as most people believe that they don't eat much food and that they eat well.

Enter Gilliam McKeith, a tough talking, no-nonsense nutritionist who starred on the UK dieting show "You Are What You Eat." This series featured a weekly story of an overweight person or family who seriously needed help changing their diet.

Often, these weighty families were in denial about the amount of food they were consuming, but also with the quality and variety.

In every episode, individuals would have their full week's diet placed on a groaning table that would almost be bending under the sheer weight of all of the unhealthy food. The fare always looked remarkably unappetizing, as the surprising amount of junk food all blended into one big beige mass.

A list would then be read out of all the food "36 pieces of white bread, 20 packets of chips, 1 tub of margarine, 14 cups of tea, etc."

You can see an excerpt at: http://www.digicast.com.au/diet-show/

Gillian would then let the shocked person know "You eat this in one week." And just about every participant would erupt into nervous laughter. Then, you could literally see the realization in their faces that their diet really was appalling and they needed to change (cue tears).

Telling people how much food to eat is too abstract, as many of us are unaware of daily food limits. It's also really hard to explain to someone that their diet is "bad" as there are so many grades of bad from: "really bad" (using heroin) to "everyone does it bad" (snacking on potato chips).

Giving people a rational, logical argument to change their diet never works.

However, by laying out the food and then later comparing it to a table of healthy food (which featured bright coloured and appealing fruits and vegetables), it became frighteningly clear to the dieter just how much change was required. Participants could so easily see just where they were going wrong.

The same is true with safety. Telling people that they need to fix up their unsafe behaviour isn't enough. You need to show them.

Another concept that is difficult for us to grasp is large numbers.

In order, to explain large numerals, we need to compare them to something more familiar.

It's much easier to explain the large weight of a forklift by saying "A fully laden forklift weighs five tonnes. That's around 3 times the weight of an average car."

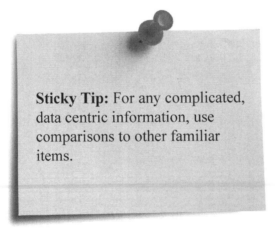

Sticky Tip: For any complicated, data centric information, use comparisons to other familiar items.

The Magic of Metaphors

Metaphors are also really handy in helping people understand and remember information. They are about understanding one thing in terms of something else. They are often sticky messages in disguise.

We all use them in our day-to-day language (often without knowing it).

For example:

Time is a thief.
Light of my life.
Rollercoaster of emotions.
I was lost in a **sea of nameless faces**.
The computer in the office was an **old dinosaur**.
Saving money provides you with a **safety net**.
Total **accident costs** can be compared to an **iceberg.**

Metaphorical thinking helps us understand others, forge empathic connections, and see meaningful relationships among information. Use metaphors to help people understand difficult information.

Sticky Tip: Include metaphors in your safety talks and writing to help bring clarity to others.

Ask Questions, Don't Tell

Research by George Lowenstein from Carnegie Mellon University found that humans don't like having an obvious gap in knowledge.

Not knowing something is like needing to sneeze, but you just can't. The knowledge gap metaphorically creates discomfort that we need to fill, as soon as we can.

It creates incompletion and many people have a high need for completion.

Interest develops from gaps in knowledge

Have you ever watched a bad movie on television? And even though you wanted to stop watching, for some reason the story line had you hooked. Despite the poor acting, you needed to know what happened to the hero in the story. That's when the "gap theory" of interest is at work.

To convince people to listen to your message, you need **to highlight some information that they didn't know they didn't know**. By asking a question about something people don't know (but they know they should), you get people's attention.

Television news programs do this really well. Their teaser promotional ads tend to focus on questions that really get your attention to find out more.

Sticky Tip: Take a note at promos used by the news programs on your television. Watch how they always ask sensational questions. Examples include "What are your teenagers doing at night?" or "Are you eating enough fruit and vegetables to avoid cancer?" Notice how they hook you in.

Our brains are literally like a big computer that sets to work to answer any question asked of it. Ask us a question and we want to know the answer.

The benefits of this are:

1. By thinking about the answer to a question, it actually makes people more likely to remember it. **The more we think about information, the more it sticks**.
2. It can be used **to change viewpoints**.
3. It can also encourage people **to take information more seriously**.
4. It creates **intrigue and stickability** because of our need for completion.

For example, people might feel a certain way about a safety process (e.g., "Sun protection - I've heard it all before"). But their beliefs about what they know about sun protection can be challenged if they are given new ways to think about it.

Even if people are initially suspicious with the information you present, they will alter their beliefs and harmoniously integrate it into their everyday perception.

This is because of an interesting psychological concept called *cognitive dissonance*. It means that we feel uncomfortable when we hold two conflicting beliefs. We are psychologically motivated to reduce any inconsistencies.

As Cialdini says in his book "Influence: The Psychology of Persuasion", humans are driven by the need to be consistent. Having two opposing viewpoints creates a sense of internal discomfort. For example, if you get someone to write down that they like to ensure that everyone around them is working safely, the next time they walk past a fellow worker doing a process incorrectly, they will be more likely to stop and correct that incorrect safety behaviour.

Just by the act of writing down that they care for the safety of others, or even by saying this publicly, makes people more likely to watch out for fellow workers. The physical discomfort of cognitive dissonance encourages people to follow through with their beliefs.

When it comes to writing safety communication, you have to break down the resistance workers are going to have that "we've heard this all before."

Create curiosity by changing your writing mindset from "What information do I need to tell people?" to "What questions do I want our employees to think about?" What knowledge gaps can you then point out to them?

This is important because the more brainpower we devote to a topic, the more likely we are to remember it.

It's also very powerful to start any talks with a question. This is because if you're doing all the talking, you're not relating to your audience. By asking questions, it's like you are in the audience's mind and you'll build rapport.

Know it alls

Polish astronomer Copernicus summed it up by saying, "To know that we know what we know, and to know that we do not know what we do not know, that is true knowledge."

While many of us want to find out what we know we don't know, the only thing that complicates the gap theory is that **we often believe we know more than we do.**

We've all experienced a team member who thinks they know everything and closes off to learning anything new.

Just asking a question won't necessarily work for them. Instead, you have to add an additional process.

To prove that a knowledge gap exists, for "know it alls" you need to:

1. Highlight what they know, then point out the gap. For example: "This is what you know. Now, here is what's missing"; then
2. Ask them publicly (either in front of the whole group or one on one) what they think is the answer to a question.

For example, the next time you're introducing a new safety theme in a safety meeting:

1. Before the actual meeting, **find out what people know about the risk before you introduce your communication**. Ask a couple of people questions about your new safety topic to gauge what people know. By undertaking audience analysis, you'll be able to make the content more relevant.
2. At the meeting, introduce your topic by providing context and enough backstory, so that people care about the gaps in their knowledge.
3. Ask people a question that you know they won't know. For example: Can you tell me what is the most common workplace injury?
4. Get people to publicly commit to an answer.
5. Encourage debate over the answer, even after you have given the correct response.

By asking people to commit to an answer in public, you're more likely to get them engaged and curious about the result. They're also less likely to be overconfident (therefore, ignoring the important lesson).

Even better, they will absorb the lesson provided that it dawns on them that other people disagree with their answer or that they are wrong.

This is because "being wrong" can work in two ways to change behaviour:

1. **Conformity** this is an incredibly strong human driver. Being wrong in public makes us all very uncomfortable.

   Our need to fit in actually makes us more open to learning the correct outcome, so we don't feel publicly humiliated again. The desire to fill a knowledge gap can be ignited, if people acknowledge that they don't know what a lot of other people do.

2. **Cognitive Dissonance** – As mentioned previously, one of the mental cues we use when we are dealing with other people is to be always consistent. There is a subtle social pressure to be congruent with your answers.

   We all believe that if you take a stand on an issue, then you must remain consistent with your beliefs. This is a very powerful technique to leverage when trying to persuade people face to face. By getting employees to publicly accept undertaking a new behaviour or understanding a new process and ideally, signing a document that says that they will do the new behaviour, their other behaviours will gradually come into alignment with their new belief.

However, it's important that you let people know they're wrong in a neutral, non-judgmental manner. Your goal is to not start a fight with employees and tell them that they're incorrect. You want to change their beliefs and avoid making them defensive.

In fact, you want to get people to think about what they don't know and to take in the new information **so that it feels like they made the decision independent of your communication.** This will also ensure that the new belief becomes embedded into their new behaviour and they are less likely to change it in the future.

Summary Tips

Remember

1. Promote and repeat one core message; it helps people avoid making bad choices by reminding them of what's important. The more we whittle down the information into a single idea, the stickier it will be.

2. Our brains process meaning before detail.

3. By providing your core concept first, you help people understand information nearly 40% more often.

4. Start your writing with your core concept or lead using the 5W's or the "What, Why, How" formula. **Alternatively, write one sentence that makes people want to know more.**

5. The brain likes hierarchy. After your summary information (or your lead – who, what, where, why, when), then explain additional information in a hierarchical manner with the most important details first down to the least important. Use the inverted pyramid method.

6. **Provide just enough information to be useful and to grab attention. Then you can provide a little more and then, a little more.**

7. Once we know something, we can't imagine not knowing it. This makes it hard for experts on certain topics to be able to think about problems from the perspective of lesser-informed people. Be aware of this.

8. Large numbers are hard to remember and understand. Where possible, compare difficult concepts and numbers to something familiar. Use metaphors.

9. **Curiosity develops from gaps in knowledge.** Leverage natural inquisitiveness by asking people questions to highlight what they do not know. Rather than approach your safety communication from a "What information do I need to tell people?" change it to "What questions do I want our employees need to think about?"

10. Get workers to publicly admit their beliefs on safety (in a positive way). Due to the physical discomfort we feel when our beliefs are inconsistent, you will be able to gradually encourage employees to display positive safety behaviour throughout the day.

PART IV

PERSUASION

Chapter 4: Get the Effective Communication System

Left Brain, Right Brain and the Whole Brain

There are three ways to communicate your safety information – by communicating to the left brain, the right brain or the whole brain.

Our left brain is rational and analytical, it loves logic.

Our right brain is instinctive, empathetic, understands context (the left brain handles what is said, while the right focuses on how it's said), nonverbal and emotional cues. **It sees the big picture**.

Left Brain

Left brain centred safety communication focuses on accident/incident statistics. It tries to get people to agree to change by looking at figures and discussing a rational argument for change. It's when you give people lots of data and information.

Right Brain

Right brain communication starts with a story about an employee being injured and shows relevant pictures (not so much the injury – but related to the story). It would talk about what happened and how that person's life was affected. It would build an emotional connection to the information.

To get your workforce to see the big picture, you need to communicate to the right side of the brain.

Knowing something doesn't necessarily translate into changing behaviour. You need to make people feel something.

Both brains

According to "In the Heart of Change" by John Kotter and Dan Cohen, the sequence of change is **SEE-FEEL-CHANGE**, not **ANALYZE-THINK-CHANGE**. Highly successful change efforts involve helping people seeing the problem or solution by feeling the emotions.

Trying to get staff to improve safety behaviour through analytical argument is like stepping through a maths equation to show workers the benefits they will get. It just misses the whole human element and what drives people.

Both sides of the brain work together - but they have different specialties.

The left hemisphere knows how to handle logic and the right hemisphere knows about the world. Put them together and you have a powerful thinking machine. Get them to work separately and life becomes well, one-sided and a little strange.

Effective safety communication appeals to both sides of the brain.

The SELLSAFE Communication Model

Now that we've gone through the important commandments of a safety campaign, what grabs attention and how to get people to remember, it's time to start putting it all together.

You're now ready to discover the eight elements to include in your safety communication that will persuade others to undertake correct safety behaviour. Later in the book, you will learn about the templates that have been created to quickly allow you to incorporate these elements.

Based on all the important concepts that you have learned so far, we are now going to put them all together into a handy mnemonic device to help you remember.

The elements are:

Simple Emotion Look Lasting
Story Authority Focus Energy

By including as many of these concepts in your safety communication, both small and large, it will help you avoid any "Curse of Knowledge" bias that you might have from being an expert on a topic.

It will also provide ways to integrate information that makes both the left and right sides of the brain work together.

Best of all, by using these elements, you will start to get traction on positive safety behaviour.

Some of the components we have discussed in previous chapters. But for those of you who have just started reading here or who need a refresher, we will provide a brief summary.

At the end of this chapter, we will workshop how to create communication using the information you have learned to date.

1. Simple

As we detailed in the previous chapter, to get attention you have to provide easy, stripped down information that contains core information. You need to write about the heart of the matter.

Humans can only learn and remember so much information at once. The more information you give people, the more they can get paralysed by it. It can even make them react irrationally or remain with the status quo.

Prioritising information by providing a clear message, rescues people from having to work out what to do next. Keep in mind that our brain likes to rely on mental shortcuts to save time and avoid being overloaded.

Essentially, you need to focus on one clear message and remove redundant information.

Supply just enough information to grab attention and then provide a little more and then, a little more.

This will ensure you create authentic messages that are easy for people to digest. It means staying on topic and avoiding any potentially damaging miscommunication.

In the next chapter, **Create**, we will take you through the process.

2. Emotion

Heartfelt messages are the way to influence change.

In the book, "The Heart of Change" by Kotter and Cohen, they discovered three steps to influence positive change.

1. **See** – Compelling, eye-catching visuals or information to highlight key problems.
2. **Feel** – Visuals that hit people at a deeper level.
3. **Change** – Emotionally charged ideas to change behaviour (remember: use surprise or fear).

Often, a lot of companies, and some safety professionals, assume this approach works:

1. **Analysis** – Give lots of detailed information and provide data, presentations, and reports.
2. **Think** – Expect the data to change people's minds.
3. **Change** – New thoughts will automatically change behaviour.

Of course, you now know that people are really bad at understanding lots of information (if you don't believe me, or have forgotten, refer back to the chapter called **Remember.** No pun intended). The more information you give people, the more overwhelmed they get and the more likely they will ignore you altogether.

There are two types of emotional content for you to work with (as detailed in the chapter, Attention): *surprise* and *fear*. Working with these types of information in your communication enables you to make your communication more *interesting*.

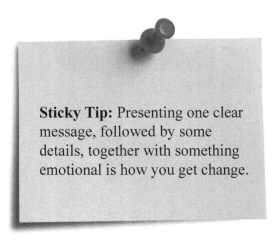

Sticky Tip: Presenting one clear message, followed by some details, together with something emotional is how you get change.

Emotionally charged events burn the experience onto our brain. Our brains have evolved to learn from emotional events so that we don't put ourselves in danger.

Feelings inspire people to act.

Think of charities. Mother Teresa said, "If I look at the mass, I will never act. If I look at the one, I will."

Researchers have found that people are more likely to give money to a charity if they hear an emotional story from one of the people the charity helps. However, if a charity just talks in general terms about who they assist (or even worse, uses a large number to describe the amount of people suffering such as "three million people in Ethiopia are without food"), sadly, they will actually discourage donations.

When it comes to safety, use emotional stories of how people's lives have been affected by poor safety practices.

As an aside, we have actually produced two ads for charities (or Community Service Announcements, CSAs, as they are correctly known).

One used Geoffrey Rush for credibility, while the other charity had the real life story of a mother and child talking about the young daughter's experience with cancer.

http://www.digicast.com.au/csa/

Watch them and think about which you found the most powerful and which one inspires you to donate.

Feel free to discuss this on our Facebook page: https://www.facebook.com/groups/484522694988251/

It's all About Me

Another way to hit the emotions is to focus your communication to attract the most important person to everyone: themselves (whether they realise that or not).

Self-interest is one of the biggest human motivators of all time. Tap into this by getting into the mindset of your target audience. While writing, keep asking yourself all the time: "So what? What's in it for me?"

It's all about getting in tune with the way your audience thinks and being in harmony with their mindset.

> Tell me quick and tell me true,
> Or else dear friend, the hell with you…
> Not how this product came to be,
> But what the damn thing does for me!
> - Anonymous

In your writing, talk in the second person and use "your" and "you". Make the person reading your material feel like *you* are talking to them.

Thankfully, when it comes to human drivers, we are biologically programmed to want to feel safe. This makes it much easier to write safety communication that matters to people.

According to Maslow's Hierarchy of Needs, once our physical needs are taken care of (food, water, shelter), what takes precedence is our desire to be safe.

This means we are programmed **to want freedom from fear, pain and danger**. We want to work in safe workplaces where we can't get hurt. We also want to make sure our loved ones are protected.

By tapping into the very essence of what drives our behaviour, safety communication campaigns can be successfully based on the strong emotional drivers to:

- Live in physical comfort with no injuries or illness,
- Protect your child from crossing the road, swimming in a deep pool and riding their bike in traffic, and
- Work safely and be protected from potential accidents.

Create your communication around these core drivers. For example:

- Do you want to risk living without your hand?
- Is the safety of your children important to you?
- How important to you is working in a safe environment?

In fact, The Shannon Company managed to tap quite effectively into our strong human desire to be safe with their advertising campaign "Homecomings" which features children waiting for their father to come home safely from work. This is a good example of how we are emotionally wired to be safe and to protect our families.

http://www.digicast.com.au/homecomings/

Later in the chapter, we will go through a case study of how Grey, an advertising agency that is best known for its social behavioural change campaigns, creates their award-winning ads.

One of the things they wait to hear in focus group testing is literally the comment "Oh, that could be me!" Once they hear that remark, Grey knows they are onto a highly effective ad.

Sticky Tip: Create your workplace safety campaign around one of these drivers:

- freedom from **fear, pain and danger**,
- **care or protection of loved ones**, and
- **comfortable** working conditions.

However, one of the things that make me uncomfortable with using a self-interest angle is that you develop a workplace where everyone is out for themselves. This can create problems that run a lot deeper than any safety communication program can solve.

As safety artists, you all know how important it is to get people to work together on safety. The worst thing you could do is encourage a Lone Ranger out to protect himself or herself at the expense of other people.

While I recommend thinking about how to write your safety communication to make it interesting for your audience, stay away from developing communication that makes it too much about an individual. This is okay for television advertising when people are watching ads in their home, but it is risky for communication in your workplace.

We see ourselves in terms of other people and groups. Evolution has taught us that it is beneficial to live in tribes, where we can share out the work of daily survival. The power of the group is a very strong influence.

As Chip and Dan Heath say in the book "Made to Stick," people also make decisions based on their group identity. In fact, changing beliefs is much easier when it occurs in a group.

Thankfully, the solution to getting people to work safely together is actually workplace communication. Discussing a safety dilemma helps forge a group identity, which in turn enhances concern for everyone's welfare. It develops group expectations and puts pressure on group members to follow them. Face to face communication is especially persuasive as it enables people to commit themselves to co-operation (Kerr et al, 2004).

If you have a strong workplace culture, then you will have a strong group identity. Group communication should focus on "we," while more individual information should be written with "you."

In a safety meeting you would use the word "we", but in an email newsletter you would personalise the information and make it more about "you", while always keeping the greater good of the group in mind. For example, you would say how at XYZ firm, it's important that *we all* use sun protection and you do this by wearing *your* sun hat and ensuring *your* workmates are wearing sunscreen, etc.

Just make sure people are accountable for their own actions. You can read more about the importance of the group and communication, in our free Workplace Safety Culture report http://www.digicast.com. au/workplace-safety-culture.

Another way to leverage the strong desire to be part of a group is by referencing a group that your workers identify with. For example, hitch your message onto an aspirational group.

We have done this successfully for a couple of clients, by encouraging workers to warm up by comparing the warm-up process to what footballers do during training and before a match (in a training video). This produces what is known as a bandwagon effect where no one wants to feel left out, so trainees will follow the information.

You can see the clip here: **http://www.digicast.com.au/warm-ups/**

This works extremely well, if you have workers that are footy mad (or whatever sport is more relevant for your organisation).

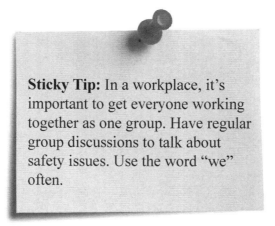

Sticky Tip: In a workplace, it's important to get everyone working together as one group. Have regular group discussions to talk about safety issues. Use the word "we" often.

Use Emotions for the Group

One thing to keep in mind when creating a safety campaign is to arouse emotions that make people want to care, rather than focus on negative emotions that blame the person for the issue.

The book "Made to Stick" talks about how in the 1980s, the American state of Texas had a big litter problem. Each year, the state spent $25 million dollars to clean up the mess and the costs were rising. A campaign was developed that appealed to the emotions and focused on "Please Don't Litter" and "Pitch In."

However, the campaign failed to work because it subtly placed personal blame and guilt.

The typical demographic of an average litterer was a pick-up driving male aged 18-35, who liked country music and sports. Authority was a bad word.

Interestingly, self-interest didn't work with these men, as there was nothing for them to gain by not rubbishing. After all, putting garbage in bins takes work and effort. Nor was giving out fines an impediment, as these guys were blatantly anti-authority.

However, Texas was able to change the behaviour of the typical litterer. By broadcasting a commercial that had two Dallas Cowboys picking up rubbish on the side of the road talking about how bad it was that a particular guy was throwing rubbish. Their main message: "Don't Mess with Texas" (this is a slogan, not a meme, as it doesn't tell you what to do). You can see it here at: http://www.digicast.com.au/texas/

Other celebrities were brought in for further ads and they all targeted macho, young males with the subtle message of "don't litter."

By using celebrities that this demographic looked up to (credibility), the ad appealed to a group identity of being a Texan, rather than a self-interest appeal of "Please don't litter".

The result was that a high 75% of Texans polled could recall the message and within a year littering had declined by a pleasing 29%. After five years, visible roadside litter in Texas decreased by 72% (and Texas is a big state!).

Research by Robert Cialdini has found that advertising which tells people that a lot of folks are doing drugs, littering or taking alcohol actually has a curious effect; it makes that behaviour appear more acceptable.

Just by labelling the behaviour as socially discouraged, but widespread, actually backfires. This is because the subtle message is that everyone is doing it (therefore, it's okay).

It's much more effective to let it be known that a bad behaviour is only being done by a small minority (or make no reference to lots of people doing it). Remember, we are drawn to being in groups and we are naturally drawn to being part of the "bandwagon."

Or, as in the case of the "Don't Mess with Texas" ad, it showed that the majority of people (especially credible footballers) all use bins, rather than litter. Only a minority litters. In fact, they quite cleverly, only referenced one person as being the litterer.

For example, Cialdini undertook a research project using this original sign to deter wood theft at a national park:

The sign showed a person stealing a piece of wood with a red circle and bar imposed over their hand.

A second sign was tested that said:

The second sign decreased theft by almost 80%.

For those interested, take a look at the environmental awareness ad, "Iron Eyes Cody" that has won many awards (it has the slogan "People start Pollution"). Cialdini has since claimed that it is a complete failure as it normalises littering by letting people know that everyone is doing it.

http://www.digicast.com.au/iron-eyes/

3. Look and Feel

As we talked about in the previous chapter, ensuring that your communication materials have a consistent look and feel is important.

By using the same colours, font style, language, message and design people will start to instantly recognise the message. This is because it creates a mental model for readers and enhances their understanding of the topic. It is actually known as a graphic identity.

In addition, it's also got to look good (if it looks amateurish, people will ignore it) and it has to follow graphic design principles to get attention (more about that in the next chapter). It also needs to include easy to understand visuals (more in the next chapter).

4. Lasting

To make a long-lasting impression, leverage advertising agency techniques and repeat your message in *multiple places, multiple times*. This was discussed in more detail in the first chapter, but it's important that you repeat your message so that people see it at least 4-7 times.

After all, the more people see a message, the more they believe it.

However, within your writing, make sure you only a repeat message up to three times. Any more than that and people will ignore it.

5. Stories

> *"Words are how we think, stories are how we link."*
> - Christina Baldwin

Stories are my favourite communication element because if done well they can be so powerful.

When I have the luxury of walking my two young daughters to school, I like to tell them a story that has something they can learn from. I don't plan my story; usually something will bubble up into my brain when I see something (for example: a type of flower that my Grandpa used to grow and then some information on that plant).

What's really interesting is that both my kids go quiet and listen! Not only that, they will often hold my hand on the way to school and say, "Mum, what stories have you got from when you were young?"

While we could be fooled to believe that stories are only for children, the truth is adults love stories, too. After all, why do you think television shows, YouTube videos and magazine articles are so popular? They're all filled with stories (and gossip!). Even the Bible is crammed with stories and metaphors, to help us comprehend information.

For ease of understanding, talk to people in 'word pictures'. Provide visual imagery, just like the previous example of the news lead about the body that was found given by Chip Le Grand.

Our right brain prefers stories. We remember stories better than if we were just told facts. They provide an emotional connection to information.

To use stories effectively, as the leader you must learn to see the relationships between one thing and another (just like metaphors). You must design stories that change the mind and behaviours of your audience.

Stories can also be used to provide examples of the impact of employees' actions on others. They are very powerful at changing inappropriate workplace behaviour. It puts knowledge into a framework that helps staff understand how they are supposed to act.

In addition to using stories to explain consequences, they can also be used for real-life examples of staff who have exhibited the right behaviour. Use stories to align staff with the behaviour that you want.

Also, stories can inspire people to improve their performance. Focus on stories that talk about those who have made it against the odds.

In our toolbox talk training course, we recommend that every safety meeting has at least one story. Some storyline examples include:

- **A positive story about an employee doing the right thing** – to reinforce the type of behaviour you want and to also provide positive feedback for the person who did the correct behaviour (which is a powerful reward in itself).

- **How another person at another company was injured** – scan safety magazines and the daily paper for workplace accidents that could have happened at your workplace. Tell the story to your team. Focus on what went wrong and ask people what they would have done to avoid the situation.

- **Get other employees to tell their story** – have other staff members contribute and tell stories about safety hurdles that were solved in their previous jobs, how they determined there was a secret danger lurking in the building and anything else relevant.

- **Start with a mystery** - Robert Cialdini did a little book research to uncover how scientists made science interesting to students. He found that it was those that started their lesson with a mystery followed by a question, that got everyone involved in the lesson, desperate to find out the answer at the end of the lecture. Think about what mystery you can incorporate into your safety theme? (That reminds me of one of my favourite Simpsons quote. When Chief Wiggum says to Bart and Ralph, "What is your fascination with my forbidden closet of mystery?").

To this day, the most popular blog article I've ever written was a story about how Paul O'Neill, a past CEO of Alcoa, transformed the company about safety. It's also an interesting story to tell your team. Read it at:

http://www.digicast.com.au/blog/bid/90674/Is-this-the-best-CEO-Safety-Speech-ever

The beauty with using stories is that you can actually start to use a whole lot of the categories in the SELLSAFE formula. It's an easy way to tick off "simple", "emotional", "authority", "focus" and of course, "story"!

What stories can you tell in your workplace? If you want a bit of extra help, read this article on how to be a story finder – How to Find the Right Stories for your Company.

http://www.digicast.com.au/blog/bid/85582/How-to-find-the-right-stories-for-your-company

If you can't think of a story use a metaphor to help explain information.

6. Authority

> "Since 95 percent of people are imitators and only 5 percent are initiators, people are persuaded more by the actions of others than by any proof we can offer."
> Cavett Robert

If I were to tell you that I have a new, wonderful safety training course that is really the best and you just have to buy it. Would you believe me? Or would you believe a fellow safety professional who told you that he's just done a great online course and you just have to do it!

Assuming you are like most people, sadly for me I would be seen as unbelievable (and annoying), while the safety professional would be found to be more credible. You'd believe the actions of the person who has already undertaken the training (unless, you knew me really well and had undertaken previous training courses of mine).

In "Influence: The Psychology of Persuasion", Robert Cialdini talks about social proof. One way humans determine what is correct is to find out what other people think is appropriate. In particular, we view behaviour as more correct in a given situation to the degree that we see others performing it.

As mentioned previously, the brain loves to rely on cues to make the processing of information easier. By seeing that people we trust, such as experts, associations, companies, celebrities and our peer group like a certain product or service gives us the reassurance that we don't need to do any further research. We can just accept their endorsement. The hard work has already been done.

An example is canned laughter in comedy shows. Despite laughter sounding corny and fake after a joke on a sitcom, research has found that canned merriment causes an audience to laugh longer and more often. Audience members will even rate the material as funnier compared to a show with no laughter track.

Social proof has been crucial to the success of eBay. Due to a rating star system of a seller's performance, buyers can see whether or not the sellers provide good service. This establishes the confidence to buy from someone they don't know, often on the other side of the world.

Likewise, an online training course that I sell through Udemy (https://www.udemy.com/supervisor-leadership-skills-for-a-safe-workplace) had better sales once I reached 40 viewers and thanks to some 5 star ratings. After all, which training course would you buy? The training course with only a handful of subscribers, or the training course with hundreds of customers and five star ratings?

Marketers use social proof extensively.

Over the last 15 years, we've used video testimonials significantly for our clients. Rather than create a promotional video that just talked about the brilliance of a particular product, we had customers give glowing reviews instead. While written testimonials are great, video testimonials are even better. **Print makes people seem less trustworthy and their message is more likely to be questioned**, unlike video which conveys more emotional impact and body language cues.

But don't think credibility is only important for selling products and services!

It's also really important when selling your safety message.

Here are some examples of when to consider including a testimonial in your safety communication:

1. **"It won't happen to me"** – Known as optimism bias, people perceive that an accident is more likely to happen to someone else, rather than themselves. Staff with optimism bias think that they're safe, so they switch off to safety messages and training. You can get around this by getting an injured worker to come in and talk about their experience (preferably that is similar in demographics like age and gender, as your workers). Select someone in a related industry or who was injured on similar equipment. Also, if you have any injured employees, get them to talk about it. It's important for their healing and for their workmates to listen. If you can't get someone to come in, purchase a video of someone telling his or her story.

2. **Show your customer experience** – This won't be relevant for all of you, but for some industries, ensuring that you make or use your equipment safely, also ensures the safety of your customers. For example, if you work for a hospital, rather than telling your employees how to lift patients safely, consider inviting a patient to explain an example of where they got hurt by poor lifting practices. Evidence shows that when it comes to inspiring workers to do a better job, your end users are surprisingly more effective in motivating people to work harder, smarter and more productively than senior management.

3. **Staff success stories** – While your end users might be powerful change agents, your own workers still have influential powers. Having your employees talk about a new safety process they developed and why, is a good motivator for other departments.

Senior management is also important to include. Look out for sites that are doing an exemplary job on safety and get them to explain how and why on camera, or in person.

Credibility as a Safety Communicator

One of the issues you often hear workers get annoyed about is fake safety information.

Inauthentic communication occurs when a safety communicator does not really care about, or understand, a new procedure or a common safety process.

While they might tell others that they care (and even convince themselves that safety is important), no one truly believes them. It's because **their words do not match their actions.** They're out of alignment with what they are saying.

Speaking and writing authentic communication boils down to one thing – the writer/speaker has to believe what they are saying.

Otherwise, the communication will come across as fake and inauthentic. Do whatever you can to ensure you feel comfortable with what you are proposing. After all, everyone will look to you as for the right way to behave.

Essentially, there are four requirements when you are creating believable safety communication:

1. Make sure you believe in the safety process.
2. When you see someone performing steps in the process incorrectly, you must pull them up and correct them.
3. You must follow those steps perfectly every time. No excuses. Otherwise, any safety communication you create will fail.
4. Follow up afterwards. This means talking to people about the new process and finding out how it's going.

These steps will provide evidence to everyone that you are serious with your safety communication. It will help them **trust you**. After all, effective leadership is based on people having confidence that the safety leader means what they say.

This is the key to creating genuine, heartfelt communication that people will believe and follow.

Stats, not possible!

Wherever you can, use statistics to give credibility to your story. After all, around 73% of people believe information if statistics are included. Okay, I made that up, but I bet I fooled you (just for the record, odd numbers are believed to be more authentic than even numbers, so if I used an even number you might not have believed me)!

Using credibility in your safety communication brings information to life and contextualises the information, so it's more human.

As mentioned previously, numbers including statistics are difficult for people to comprehend. **It's scale and context that make statistics more real for people.**

Rather than give people numbers, try and show the information in a visual way.

For example, in Kotter and Cohen's book, "The Heart of Change", there is a story told by Jon Stegner about how the company he worked for was wasting lots of money purchasing safety gloves.

Seeing an opportunity to drive down the costs by 2%, which would work out to be around $5 billion over five years, he tried to convince management to do something about it. Yet, no one listened.

It was only after an intern completed a small study that they discovered that 424 different types of gloves were being purchased, as every factory site had their own supplier with their own price. Gloves varied from $5 to $17 a pair, but when you are buying a lot of gloves, the price disparity makes a huge difference to total costs.

To get the point across in a concrete way, all 424 types of safety gloves were labelled with the price and the name of the factory using it. They were then all showcased on the company boardroom and division presidents were invited to visit the display.

Their mouths literally dropped, rendering them speechless (surprise!) when they saw their expensive boardroom table stacked with gloves displaying a bewildering array of styles and prices.

The gloves became part of a travelling road show and Jon Stegner got people on board with cost cutting.

How can you incorporate a visual display to bring more credibility to your safety communication?

7. Focus

One of the reasons why you see so much bad safety communication is because the majority of communication tries to be all things to all people.

There is no clear goal as to what the communication is trying to achieve, rather it brings in lots of different messages in the hope that one of them will stick. This only creates inconsistency and confusion.

In order to write **targeted and relevant communication**, think clearly about:

1. **What is the goal of the communication?** Start by writing out your objective. Eg: To increase awareness about the new safety process.
2. **What is your one main message?** What is the main focus of the communication?

3. **Who is your audience?** Ensure your writing focuses on your target market.

4. **Are you focused?** When you are writing, are you focusing on the job at hand? Or are you worried about that report you have to get out on Monday? One technique I use is to actually write down my intention of what I want to achieve before I begin a new project. This might happen a few times during the day, but it ensures I give it my full attention. For example, I wrote an intention on a sticky note of what I wanted this book to achieve and I looked at it before I started writing each day. This can also be your objective that you regularly refer to as you write.

In the next chapter, you will learn tips on how to focus on these areas, which are then brought together in the SELLSAFE Communication template.

8. Energy (Action)

Advertisers know that there is no point in just promoting the virtues of a new product.

What advertisers must inspire you to do is take action – this is most commonly in the form of making a purchase.

However, in safety communication, it's all about getting people to think about the safety issue and make changes to their behaviour.

This is what Grey advertising does so well. They always end with a safety meme that gets people to question their behaviour to elicit change, such as "Would you do what you ask your Workers to do?"

In fact, as you'll read at the end of this chapter, Grey always undertakes post-advertising research to test the viewer took action as a result of the ad. A score between **40-50% is required.**

This is a critical component of safety communication. In particular, if you do use fear in your communication, you must clearly specify the steps to avoid the danger (that are easy to perform), so as not to lose people.

Getting the viewer to act and expend some sort of energy on the message is crucial.

We'll go through how to do add action statements to the end of your communication in the next chapter, **Create**.

Two-Way Action

In addition, to telling people what you want them to do, make sure the action is two-way. Give people the opportunity to let you know that the action required is or isn't working.

Over time, people will end up changing a safety process. What they were taught in their procedure can end up morphing into different safety behaviours. The risk is that the new method is unsafe.

Empower your employees by allowing them to inform you when they want to change a process. However, you must stress that you need to check that it is both safe and efficient.

To make this work effectively, it's really important that employees know that:

- you will not be angry, if their new way is unsafe or less productive, and
- you are still pleased that they took the initiative to consider new solutions.

The important thing is that everyone knows to never change any safety processes without approval.

As Peter Drucker was quoted saying in the book, "Inside Drucker's Brain," improvement and abandonment are two sides of the same coin. Companies must always be getting rid of obsolete processes and empowering their workers to constantly look for ways to rid the company of unnecessary waste.

To improve one must abandon what doesn't work and do what works better.

However, as humans have a tendency to change their processes over time without knowing it (and make the task less efficient), ensure you audit against work instructions on a regular basis.

But now, time for some examples.

Safety Communication Workshop

Time for some fun!

To help you understand how to use all of the tips from the previous chapters, I've assembled various statements on sun protection. In the second example, I include a range of techniques to change how the information is presented to make it more powerful.

Example 1:

Protection from ultraviolet (UV) radiation is important all year round, not just during the summer or at the beach. UV rays from the sun can reach you on cloudy and hazy days, as well as bright and sunny days. UV rays also reflect off surfaces like water, cement, sand, and snow.

The hours between 10 a.m. and 4 p.m. daylight savings time (11 a.m. to 2 p.m. standard time) are the most hazardous for UV exposure.

A lot of people think that applying a sunscreen provides 100% sun protection. But it doesn't. While higher numbers of SPF factor mean more protection, many people are confused by the SPF scale. SPF 15 sunscreens filter out about 93% of UVB rays, while SPF 30 sunscreens filter out about 97%, SPF 50 sunscreens about 98%, and SPF 100 about 99%. The higher the SPF, the smaller the difference in actual sun protection. Sunscreen doesn't last all day and needs to be reapplied. No sunscreen protects you completely. Regardless of the SPF, sunscreen should be reapplied often every 2-4 hours for maximum protection.

Never use sunscreen to prolong your time in the sun. Even with proper sunscreen use, some rays get through, which is why using other forms of sun protection are also important.

Even when wearing sunscreen, you must include other forms of sun protection such as wearing a broad-brimmed hat, sunglasses and being in the shade when you can.

Loose-fitting long-sleeved shirts and long pants made from tightly woven fabric offer the best protection from the sun's UV rays. Darker colours may offer more protection than lighter colours. Some clothing certified under international standards comes with information on its Ultraviolet Protection Factor (UPF), which tells you how much protection you can expect to get from that article of clothing.

Let's take out the important information to help us work out the core message.

What do you think is surprising or unusual? Wearing sunblock does not protect you from the sun. The higher the sun protection factor, the smaller the difference in actual sun protection.

What's the Lead? When working outdoors, it is important to avoid sun exposure, so to minimize your chance of getting skin cancer (why) always ensure you (who) have adequate sun protection. Wear a high sun protection sun block (re-apply every 2-4 hours (what), a wide-brimmed hat, long sleeves and work in the shade wherever you can (where).

What's the Main Message: How to Protect your Health by Limiting Sun Exposure or Why Sunscreen Alone Won't Protect you from Skin Cancer.

Question for Staff: Does a high sun protection factor sunblock of 50 protect you from sunburn all day compared to a sunblock of 30?

Let's rearrange it:

Why Sunscreen alone won't protect you from Skin Cancer

When working outdoors, it's important to avoid sun exposure to minimize your chances of skin cancer. Sunscreen does not provide 100% sun protection. Even high SPF 50 sunscreen only filters out 98% of the sun's rays.

To ensure you have adequate sun protection, you need a number of strategies, to provide full sun protection:

1. Apply a high sun protection sun block (re-apply every 2-4 hours).
2. Wear a wide-brimmed hat.
3. Wear loose-fitting long-sleeved shirts and long pants/skirts made from tightly woven fabric which offer the best protection from the sun's UV rays.
4. Work in the shade as much as you can.

Never use sunscreen to prolong your time in the sun. Even with proper sunscreen use, some rays get through, which is why using other forms of sun protection are also important.

Get sun smart and always use multiple forms of sun protection.

To learn more about sun protection, visit Sam in the safety department to get your handy sunsmart information card with a free sunscreen sample.

In the next example, we will go through the importance of three points of contact for stairs. Many of you have probably experienced trying to give this message, only to be ignored by others because "they've heard it all before."

The format is more suitable for a newsletter or your intranet.

Figure 1: Could your stairs be dangerous?

Which is more dangerous – stairs or sharks?

Did you know that you're more likely to injure yourself by falling down stairs than by being attacked by a shark?

Stairs represent a serious injury risk and are one of the most common causes of injury at the workplace and in the home.

And here at XYZ, they represent our most common form of injury. To keep safe on stairs:

- Always have three points of contact.
- Avoid talking, reading or texting while going up or down the stairs. Pay attention to each step.
- Ensure stairs are clear and are never used to store items.

Remember, just by being aware that stairs are dangerous, paying more attention and using three points of contact will keep you safe from the dangerous sharks, I mean stairs.

If you have not had three points of contact training or you are confused by what that means, please contact me at john@xyz.com.

Analysis of Three Points of Contact (Shark on Stairs)

Headline – uses a question format to get people thinking about something unusual. Most of us believe that shark attacks are common (they're not). So to break down people's guessing machines, we're surprising them with the fact that stairs (which we all use every day) are more dangerous than you would normally think.

Image – Stairs and a shark are combined to give an unusual picture that grabs attention. Humans get emotional about sharks (i.e., fear), so the picture helps convey the danger inherent in stairs. The first image has a caption underneath to grab attention and hook the reader with the main message, while the second photo integrates text and pictures to explain the process.

Sub-headline – A further bolded heading is used, to get people thinking again with a further question. Remember, it's important to let people know what they don't know, to get attention.

Main copy – The most important, general information is specified first and then the steps to avoid the danger are detailed in bullet points (remember, if you do use fear, give clear steps on how to avoid the threat). The layout makes use of white space and includes a further image to fully explain what three steps of contact are about. Humour is added for further engagement (and to lessen the fear element).

Call to action – The story is wrapped up with a call to contact John if there is any confusion or to organise training.

How an Advertising Agency Creates a Safety Campaign

Advertising agencies are blessed with big budgets and resources to create highly effective safety campaigns that have contributed to positive social behavioural change on a massive scale.

One such agency is Grey. They are a worldwide agency responsible for some of the most highly effective public safety campaigns over the last 25 years. And that's just in Melbourne.

Grey undertakes advertising campaigns for both WorkSafe (the Victorian workplace safety authority) and the Transport Authority Commission (which provides compensation to transport accident victims).

To find out more about the process Grey use to produce their ads, I talked to David Dumas, a Grey business director with over 30 years advertising experience, who looks after the accounts for both clients.

Back in 2005, when Grey just started to produce ads for WorkSafe, the wider community saw safety as a hindrance that put a handbrake on productivity. Now, it is viewed as a beneficial objective for companies.

Over the years, the many advertising campaigns that WorkSafe has broadcast have been found to positively increase the attitudes of Victorians towards workplace safety. During this time, there has been a reduction of workplace injury claims (from 9.5 per million hours of work down to 7.7).

Every ad Grey produces uses a combination of three elements: **Emotion, Enforcement, and Education**.

Each commercial draws attention by:

- having a story that is relatable,
- a surprise (which can often be fear based),
- clear steps on how to avoid the accident, and
- a closing with an informative safety slogan or meme (action-based).

Grey also ensures that they only focus on one clear message at a time per campaign.

For example, take a look at this ad that was aimed at supervisors that ended with "Would you do what you ask your Workers to do?"

http://www.digicast.com.au/workers/

To create these award-winning ads, the process begins once Grey receives information (or a brief) from WorkSafe on a problem that needs to be solved.

One of the issues WorkSafe had uncovered was that young workers were more likely to be injured in the workplace.

Staying silent at work actually contributed to young people being injured. WorkSafe wanted to create a campaign that would let young people be more aware of workplace risks.

To start the process, Grey:

1. **Undertook initial research** into the attitudes and behaviours of young people aged 16 – 24 years by undertaking focus groups with this age bracket. What they found was that young people didn't even believe that work was dangerous. To consider that workplaces were unsafe, the participants believed that they needed a "jolt" to take notice. Ads from around the world were

played to gauge reactions about the type of "shock" tactics required.

2. **Concept creation** – Once knowledge was gathered on how young people think about workplace safety, Grey was then able to craft the core message. For the young worker campaign it was: "If you're not sure, ask." Then, around 4-5 very different television ad concepts were created (all included different variations of the three elements: Emotion, Enforcement and Education). Different types of slogans were also produced.

3. **Advertising Testing** – More focus groups were recruited to test the 4-5 different ad concepts. Every detail of the ad was discussed. Usually one ad is a clear winner or elements from a couple of different ads are re-used and combined. The winning ad needs to emotionally move people and the comment that really lets Grey know that they are on the right track is: "That could be me!" (Bingo!)

4. **Pre-advertising testing** – The winning ad was then put into production. After it had been edited together, it went through another round of focus group testing with the right demographic. If there was any confusion on any aspect of the ad, it was redone. Finally, after two stages of research testing, the ad was deemed ready to go live.

5. **Post-advertising testing** – After the advertising was broadcast, monthly telephone surveys were undertaken to ask people whether they had seen the "Young Worker ad." This ad received a very high 89% in prompted awareness tracking. A further question was asked to check whether the message was understood and also whether the viewer took action as a result. A stunning 60% of workers / supervisors said they did something as a result of the campaign.

Worksafe advertising campaigns always include "lots of different tentacles" to promote their messages in multiple places and multiple times by including outdoor, digital and radio ads, as well as public relations.

Without realizing it, Grey uses the eight **SELLSAFE** elements to change safety behaviour. They always include:

1. A **S**imple message,
2. **E**motional content,
3. A **L**ook that is consistent throughout the company.
4. **L**asting communication that's shown "multiple times, in multiple places".
5. A **S**tory that weaves all of the elements together
6. **A**uthority (having credibility by showing how the accident can happen to an everyday person),
7. **F**ocus which is extreme in detail to ensure the suitability to the right audience
8. **E**nergy, a clear action is always specified.

What I like is how every ad ends with a clear call to action on what people need to do to avoid the safety threat.

This is such an important step to include if you use fear in an ad. Providing important steps to avoid the danger ensures you do not lose the viewer.

Persuasion

Summary Tips

When creating any form of safety communication, use these eight elements to help drive your message through and to engage:

S – Simple – Make sure you communicate a clear message, avoid overwhelming with information.

E – Emotional – This helps people to care and done well can ensure they remember the message. Consider using emotions such as fear and surprise, in order to make information interesting.

L – Look and Feel – Make sure your communication materials all have the same consistent look and feel. Ensure they follow great design practices that grab attention

L – Lasting – Remember, your message must appear in multiple places, multiple times. The more people see your message, the more they will remember and believe it.

S – Story – Stories help us understand information and to create emotional connections. Remember to also use metaphors to help explain information.

A – Authority – People believe authority figures and will look at the actions of others to determine their own. Consider using other people to tell their story. Also, ensure that you include real-life examples that people can relate to.

F – Focus – Ensure your clear message is very focused on your audience. You must consider the best way to communicate with them.

E – Energy – The crucial part of every communication piece you create. What do you want people to do? What is your call-to-action?

PART V

CREATE

Chapter 5: Transform your Safety Communication

> *"Think like an artist, execute like a Samurai."*
> - Jon Wuebben

Congratulations! You've now read all the theory behind how to change behaviour through safety communication.

It's now time to put it all together. Well, almost.

There is one more important concept to understand before you get started.

The Power of Visuals

You might have noticed, but visual content is dominating the social media landscape. Facebook, Pinterest, Instagram and YouTube are all visually driven. And they're booming.

In 1991, Roper Starch Worldwide found that in the "read most" category, newspaper ads that included photos were scored 60 percent (60%) higher than those with no photos.

But what's this got to do with being a safety artist, you ask?

The visual revolution doesn't just stop at just social media and newspapers. It's infiltrating all levels of communication. The need to use more visuals to engage your workforce is building in momentum.

And it makes perfect sense.

According to John Medina, the author of "Brain Rules," we are more likely to recall visual information and we are amazing at remembering pictures. Recognition soars with pictures. In fact, recognition almost doubles for a picture compared to text.

People will only remember ten percent (10%) of what you say 72 hours later.

Figure 1: Only 44% of people recognised the word chips, but 84% recognised the picture.

However, if you add a picture, it goes up to thirty-five (35%) and if you add both a picture and word together it increases to a very high sixty-five (65%).

118

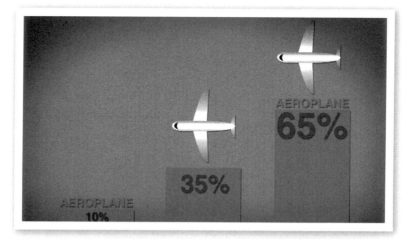

Figure 2: By including a picture with a visual, you help people remember the information for longer.

When using images in your communication to explain how a process works, try to integrate the text into the picture. This is more effective than using words alone or having the text under a photo.

Figure 3: To really help people understand and remember information, integrate text with a corresponding picture.

Research has found that when a reader has to keep switching between the graphic and its description, the brain has to work harder on what's not important. This usually means the reader misses what the picture is about.

Remember, the brain likes to use shortcuts when processing information so that it doesn't get overwhelmed. It's important that the mental bandwidth in the brain is dedicated to making sense of the topic, not how the topic is presented.

However, there is just one area that complicates this a little bit.

Captions under photos are highly effective in drawing attention. Always use a caption under a photo when the picture is not being used to explain a process.

If the photo is for teaching purposes, you need to integrate the text onto the photo for clear instructions. This is because people's eyes are always drawn to a photo and then the caption. However, there is one important thing to remember with captions:

Captions must always include your core message.

Reading Hurts your Brain

Another area that the brain has trouble with is reading. Reading is inefficient, as we have to identify certain features in the letters to be able to read them. The brain sees each letter in a word as an individual picture. This takes take time to read. Lots of reading effectively chokes your brain (see the animation below on how the brain reads).

http://www.digicast.com.au/visuals/

Visually rich presentations keep the eyes busy and therefore, the brain more active and alert to learn information. The right brain prefers visuals and can process pictures 60,000 times faster than the verbal brain can process words.

In the book "Cashvertising," author Drew Eric Whitman writes that the less imagery you convey, the less your message occupies the reader's brain, meaning that you're less likely to influence them.

When creating a safety campaign, always include a photo or drawing that best represents your information. Always.

Sticky Tip: When using images to teach a process, integrate text into the picture. If the photo is not for teaching purposes, then use a caption that includes your core message.

Make it Concrete

Remember when you were at school and you were taught subtraction?

A lot of kids find it difficult to understand because it's so abstract. This means it is difficult to compare it to something we already know and understand.

In the book "Outliers," author Malcolm Gladwell explains that Asian children learn math faster than Western children because there is a big difference in the number-naming systems between Western and Asian languages. In English, we say thirteen, fourteen, fifteen, rather than the expected threeteen, fourteen and fiveteen. Likewise, we use fifty and twenty which sounds like 5 and 2, but not quite.

Asian languages have a logical counting system: eleven is ten-one. Twelve is ten-two. The difference means that Asian children learn to count much faster than Western children. By the age of five, American children are already a year behind their Asian counterparts in basic math skills.

The Asian number naming system is concrete and easy to learn, while our number system is abstract.

Likewise, when Asian students learn math, they are given questions such as "You had 50 yen and you bought a notebook for 30 yen, how much money do you have?" Abstract mathematical concepts are taught by emphasizing the familiar. In Western countries, kids are asked, "What is 50-30?"

Great teachers and marketers make difficult things easy to understand by using concrete language, visuals and clear comparisons. They make difficult information concrete, not abstract.

Concrete ideas are easy to remember.

Photos and video content can often be useful to explain difficult information.

When I studied biology at school, I really struggled to make sense of how the body worked from textbooks and diagrams. It was only when I was able to see 3D animations that actually showed how the human body operates that my mind could fully grasp the abstract information.

Take a look at this 3D medical animation we created to demonstrate how the spine works for CSR Viridian:

http://www.digicast.com.au/injury

Here are some visual ideas to get you thinking:

1. Infographics

Infographics are graphic visual representations of information, data or knowledge intended to present complex information quickly and clearly. They can improve understanding by using graphics to enhance the visual system's ability to see patterns and trends.

They're also a handy way to explain complicated data.

For example, this is an excerpt from an infographic we created for a client. Their original document included six columns of detailed information about their Health and Safety Management System that was over three pages.

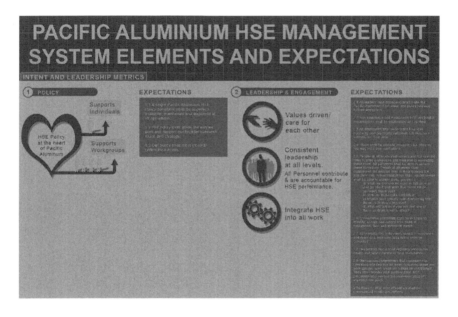

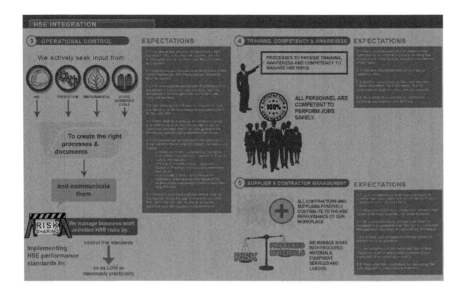

In the words of Alistair Camm, the HSE Manager for Pacific Aluminium: *"The infographic has been used as a summary of the management system elements in the corporate HSE system manual including intent and expectations. Reactions to the infographic have been positive – one of the Safety Advisers was really taken with the heart symbol for the policy. He said: "**It really gets to what we believe about the importance of the policy and making that link to hearts and minds**." The concept of infographic was also utilised by Rio Tinto Corporate HSEC as an idea in developing a similar style document – they, however, used photos of Rio Tinto workers rather than infographics – it is not as effective in getting the message across in such a succinct manner".*

Infographics are ideal for whenever you have lots of information, statistics, or a complicated process. They're very visually appealing and can make difficult information less confrontational.

Many of you have probably created flowcharts and other diagrams, which are a type of infographic.

They can also be used to provide visual instructions on how to do something. For example, below you can see two ways of telling people to drink water. Which one do you find more effective?

There are also some free programs around to create infographics (it's just tricky to find what you need). Take a look at:

http://piktochart.com/

http://infogr.am/

http://www.easel.ly/

http://www.gliffy.com (flowcharts)

Otherwise, look at getting them custom made for your workplace.

2. Graphic Icons

Years ago, I sat on a panel to judge the website design presentations for a new Government health website. One of the striking submissions had every section of the website defined by a cute little icon. It tested really well in market research focus groups because people could quickly grasp what the graphic represented.

This is a nifty little idea for companies.

Whenever you have any confusion with safety processes, particularly if you have a lot of them, it's a good idea to incorporate more visuals.

For example, a company that we worked with had five hazards that were quite complicated to delineate in a training manual. It was often difficult for readers to easily identify which hazards the document was discussing.

Our solution was to produce a graphic logo for each of the five hazards to make it easier for readers to instantly recognize the hazard being represented.

Now, every safety document that is written has an image in the top right header that identifies which hazards are discussed in the document.

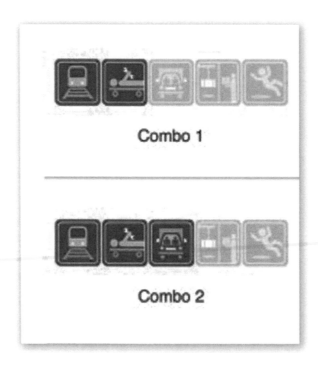

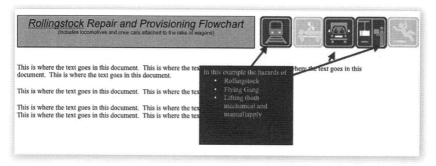

These icons are also used in other workplace communication from posters to emails.

3. Posters

The majority of you would have safety posters of some sort in your organisation. A lot of companies buy off the shelf posters, but if you are designing a safety campaign, consider designing and printing your own. This will ensure that you meet one of the golden rules of marketing – **keeping all your materials consistent**.

They are a great way to remind people of your message. For greater impact include:

- A person and their face (humans love looking at other people).
- A well-written safety slogan or meme.
- An image that matches the slogan.

Here's an example of some posters we made for the Gypsum Board Manufacturers of Australasia. Note the same consistent look and feel.

Other important visual examples that do not need further information include:

- Magnets, key rings.
- Screen savers.
- Message on security passes.
- SMS's.
- Social media (Facebook, Twitter).

4. Video

As mentioned previously, emotional imagery is more likely to get people to change. For really strong emotional impact, consider including video content.

According to US author and management consultant, Joseph Grenny, a study found that by showing video footage of a child picking up and eating food that had been spilt on a restaurant floor, motivated lazy fast food employees to clean up the floor immediately after a spill.

Yet, it's not just emotional video content that is powerful. "How to" videos are one of the most popular types of video on YouTube. And it's no wonder.

It's very easy to watch short training videos to help you learn how to use your new pressure cooker or how to set up a new cloud server system for your business (yes, I have a variety of interests).

Anything technical or complicated can be more effectively explained in a video. It's also another way of making data easier to understand. While animations are a brilliant way to illustrate how things actually work that are normally difficult to see (such as inside the body or a car engine).

There are three ways for you to produce videos:

- Create your own videos to show how machinery works, undertake safety stories, or show a safety procedure. Use programs such as:

    - http://www.animoto.com (add photos to quickly make an animated slideshow).
    -
    - http://www.techsmith.com/jing.html (capture video and screen shots and annotate).
    -
    - http://www.soundeffectsforfree.com/ (add sound to get the auditory channel open to learning).

- Purchase off-the-shelf training videos or animations that highlight what it is you are talking about. Look for safety stories that can be incorporated into your safety meetings.

- If you can't get what you need or you need your videos to be high quality, get them made professionally through a production house.

Now that you know how important it is to include visuals in your campaign, let's go into the nuts and bolts of how to put your campaign together.

Summary Tips

Create

Humans recall and understand visual information much faster than words.

Visuals engage your audience and help people to instantly understand information sooner and more thoroughly.

Whenever you are having difficulty explaining what you mean through words, consider using pictures, diagrams, flowcharts, photos or even, video. Remember the saying: "A picture is worth a thousand words."

Chapter 6: How to Make Your Safety Writing Engaging

Forget Your English Teacher

If you're like most people, you probably feel uncomfortable with writing. Hours are spent painfully trying to be grammatically correct, while an imaginary English teacher appears on your shoulder telling you: "That's not right, do it again!"

I'll tell you a little secret – writing long sentences with big words is for people who want everyone to think they're smart (i.e., academics)! The irony is that really intelligent people write in short, concise sentences (they don't need to "trick" people into believing that they're smart). Writing brief sentences that are quick to understand is the way to go.

Remember this: the longer the sentence, the more effort it will take to read it (and reading hurts the brain). Long sentences mean long thoughts (which is often a giveaway that the writer isn't even clear themselves on what they're trying to say).

> *"Please excuse the long letter - I didn't have time to write a short one."* - Ralph Waldo Emerson

When I used to write market research reports, we were taught to write simply and succinctly. The truth was that no marketing manager was going to care about our impressive vocabulary. They just wanted to know the results of their study – and they wanted to know fast. Impressing them with lots of statistics and recounting of the data would have lost them in an instant.

Actually, that reminds me of a story. I once worked with a colleague, who had a really weird habit of including one obscure word per report to impress the marketing director. If my colleague was asked what the word meant, he would bathe in the afterglow of sounding really smart for days. We'd have to hear constant snide remarks that Craig the marketing director wasn't as smart as him. I didn't work for that company for long.

But I digress.

By clearly explaining the data and what it meant (and showing lots of visual charts), we were taught how to write an interesting market research report (some people might not believe that's possible, but that's not entirely true).

No one has the time or even inclination to wade through complicated sentences. Even doctors, who people see as highly intelligent, prefer to read text with short words.

Ignore what your English teacher told you. Plain English connects with everyone at all levels. Of course, if your language is too difficult, people might think that you're trying to talk above them. This will only stop your communication being accepted by others.

Consider this official permission to write in a conversational and friendly manner. Ignore structure and formal writing requirements. Write like you speak. Use lots of short words and sentences.

The only people you need to write for are your employees, not an imaginary teacher. You can even start your sentences with "and"

and "but." For some reason, I like to do it a lot, because it feels so liberating. And, you can, too!

How to get into a Writing Mood

If you are nervous about writing, then you need to set up the right environment so it's more conducive.

It's really important that you are able **to focus** on the task at hand. To help you get into a writing mood:

1. **Get rid of distractions** (turn off email, turn off your mobile, shut your door). Clear your mind.

2. **Do any little jobs** that are going to be at that back of your mind to do (e.g., pay bills, feed your cat or put the cat outside if they like to bug you while working. My cat likes to meow a lot when I work from home, but only when I'm on the phone).

3. **Grab a coffee, water, or tea before you begin** (otherwise, you'll be immediately thirsty once you start).

4. **Ensure your desk is clean and tidy**. Might even be time to give it a little dusting before you begin.

5. **Get out your employee or contractor persona photo** – Place this where you can see it. More on this later, but it will keep you focused on who you are writing to.

6. **Write like your speak**. Ignore any grammar rules. English teacher, be gone!

7. **The first 5-10 minutes are the hardest.** Get over this hump and you'll be more focused (this does not mean celebrate by getting another drink).

8. **Keep going and write solidly for 60-90 minutes**. Gary Halbert, a famous copywriter once said in one of his newsletters that the number one way to banish writer's block is to "write, write, write!" Don't worry about grammar rules. Just get everything you need to say. Edit later. Take a break at 90 minutes by going for a short walk, even if to the kitchen to get that second cup of coffee.

9. **Have chewing gum** – I get surprisingly hungry when I need to write. I'm not hungry at all, just in need of a distraction. Chewing gum is great because it actually helps you think better (no joke!) and you avoid putting on weight. It also keeps your teeth clean.

10. **Believe that you can do it**.

Okay, now that you're sitting at your desk excited and ready to go, what do you start doing?

How to Start Writing your Safety Communication

There are four important steps that you need to be clear on before you even start writing your safety campaign. This will help you really keep **focused** on what you are trying to achieve, which is an important component of the SELLSAFE communication philosophy.

These components are included in your **SELLSAFE** template, which will make it easier for you to remember. You may wish to refer to that now electronically, as well as the *Employee Persona* template (at the back of the book).

Start your safety communication by getting clear on:

1. Your Objective
2. Your Target Audience
3. Research
4. Media

Let's look at all of these in more detail.

1. Start with your Objective

First of all, you need to get clear on what outcome you want to accomplish. What's your intention?

A lot of ineffective safety communication tries to be all things to all people with no clear focus.

Think of your safety campaign as an experiment in cause and effect. If you want to get a great effect, then you need to really work out how to "cause" that.

To focus your mind, figure out the objective of your safety campaign.

For example:

> *To educate our staff on how to safely operate a forklift.*
> *To increase awareness of the five hazards at our construction site.*
> *To reduce hand injuries at the glass cutting table.*

Being clear on what you are trying to achieve will avoid any miscommunication. It will also enable you to create authentic and relevant communication.

Write your objective down and refer to it often.

2. Define your Target Audience

Next, it's important to write down whom you are writing for.

While you could write down some basic demographic information such as:

Male glass workers aged 18 – 30 years
Female office workers aged 25 – 45 years
Males and females on the Board

There is also another more visual way.

It's called a **buyer persona**.

In marketing, buyer personas are fictional representations of ideal customers. They are based on real data about customer demographics and purchase behaviour, along with educated speculation about their personal histories, motivations, and concerns.

Marketers use them to get out of their own mindset and into the mindset of what their customer wants.

This is an important process, as so often, we write about what we think our readers need to know without contemplating how and what they need to do or whether it's even going to interest them.

Of course, buyer personas aren't quite relevant to you, but if we call them an *employee or contractor persona*, then they will get more real.

At the back of the book is an image of what it looks like. You will also find the editable PowerPoint template in your download pack.

Essentially, you need to have a good understanding about the types of people you need to persuade in your safety communication. This could vary greatly, depending upon the size of your company.

If you can, start by looking at the demographics (e.g., gender, age, occupation, education level) of staff from any employee satisfaction surveys that your company runs.

Otherwise, talk to HR and ask them what types of different demographics exist. If none of those are relevant to your company, then you probably have a pretty good feel for the types of groups just by who you talk to in any safety meetings or training.

The important elements:

- **Always include an image**. It does not have to be a real staff member, scan free photo websites and find a good representation.

- **Give them a name**. The more it represents what they're like (in a nice way), the better. For example, I call my safety buyer persona, Safety Sam, (he's a very handsome, clever safety professional who enjoys reading what I say and loves giving me lots of work and referrals). My female safety customer is also highly intelligent and we have great chats over coffee.

- **Write down what is important to them**, how they think, challenges in communicating with them, their opinion on safety and any possible objections.

- **Preferred communication format** – what's the best way to get their attention - SMS, email, face to face or all of them?

By writing down your employee or contractor persona, you will start to be clearer on how to write to them and their triggers.

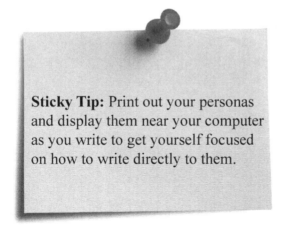

Sticky Tip: Print out your personas and display them near your computer as you write to get yourself focused on how to write directly to them.

3. Doing your Research

Now that you have your objective and a photo/bio of your employee persona, it's time to start researching your topic.

Even if you know the topic well, doing some research gets you into learning mode and will hopefully get you into the perspective of what your employees need to learn.

To start writing a new safety campaign or article:

1. **Talk to employees or contractors about the topic** – Find out what people know and want to know. Work out the gaps in knowledge, so you can formulate the right questions to engage. By getting worker input, they will have more ownership and interest in the topic.

2. **Do some web research** (type in the topic you want to educate people about into Google and use Google Scholar if you are looking for research information). Visit any relevant government safety websites for your region. Pull together any documents or books you have on the topic. Research can be a pretty quick process. Aim to find the best information you can within 15 minutes. Remember, the more you know, the harder it is to educate those who don't know.

3. **What information did you find surprising?** Don't overthink this. Write down whatever comes up first. The more you think, the more you will get bogged down in detail and lose the lead. What do you believe your employees wouldn't know? What's the uncommon sense? For example: Sunscreen does not offer 100% sun protection. Test with colleagues to see if they find the information surprising.

4. **Write your lead.** Write a summary that includes the 5 W's – what, when, why, where and who. You don't have to use this, but it's a great way to focus yourself on the information. Continually ask yourself "So what? What's interesting about this information" and cut out what you don't need.

5. **Write the second paragraph.** Provide further details in the next paragraph, organising the most important information down to the least. Use the inverted pyramid concept. Write the information as if you're writing to a friend. Use dot points to put in step by step instructions. Make sure you have included a "why" that explains why the safety message is important, as this will help people connect to the information.

6. **End with a call to action.** What do you want people to do with the information? Do you want them to call you for more? Sign up for extra training? Always end with what you want people to do.

After you have written your first draft, read it and ask yourself: what's the main message? Is there one clear message or several?

If you can't decipher the message, read it out loud to someone else and ask them what is the main message they took out of it? This is quite an interesting process and will really test how well you've communicated one solid idea.

Once you've worked out your core message, promote and repeat (maximum of three times; after that people will ignore it) and re-write the article to ensure it is focused on that one message. This can also form part of your headline.

If you can, leave your writing for at least a day. Do any editing a day after.

Use this writing to form an article and to guide all of your other communication pieces (i.e., content for poster, speech, information for report, etc.).

After you have written and published your safety communication, talk about the content with employees to make sure that they understood. If there are still gaps, create additional content on the topic.

4. Select your Media

Now that you have a good outline of your message and the related content, it's time to figure out how to best present it.

If you're writing for a monthly safety theme, it's important to use as many different mediums as you can to grab attention. After all, you've got about four weeks or 20 workdays to get the message through in **multiple places and multiple formats**. You will need to put together a schedule of publishing, but for now, work out your core media (refer to the **Communication Schedule** template).

For example, for the manual handling training awareness campaign we created for the Gypsum Board Manufacturers of Australasia (GBMA), we started with creating a 20 minute training video.

After that was completed and we had created a plasterboard character (or graphic identity) that resembled a typical plasterboard worker, we used that image in the video, posters, training manuals and workbooks. The additional elements were introduced after the main video was showcased.

3 Proven Headlines to Use

Now, we're getting into the fun part. If you're writing for a newsletter or report then you really must have an attention-grabbing headline. So many safety professionals use headlines that are drab and are more likely to turn people away.

Headlines are considered so important in marketing that copywriters will spend 80% of their time tweaking their headline. In fact, ad agencies will often write 50 headlines, to only choose one.

A good headline will grab attention and encourage the reader to continue reading the article. In advertising, headlines are often tested to see which one sells more.

Before you groan that you don't have that much time to write your headline, don't worry. The good news is, as headlines have been so extensively tested (since the 1920s), there are some foolproof formulas for you to use.

Your headline must possess:

1. Self-interest (your core message, get to the heart).
2. Curiosity (new information).
3. An easy way to accomplish a task.

Essentially, there are three safety communication headlines that I'd recommend:

1. **How To, Here's How or Learn How To** – When you start your headline with these words, it actually forces you to focus your attention on what needs to be written. It also makes it very clear for readers to know exactly what is in the content. For example:

    How to Avoid Getting Skin Cancer When Working Outside

    Here's How to Protect yourself from Sun Damage

Learn How to Apply Sunscreen the Right Way

2. **Ask a Question** – This forces people to consider what they know. As the brain is an answer-seeking device, it really wants to read more to find out what it doesn't know. One of the most highly regarded headlines in advertising is "Do You Make These Mistakes in English?" I used this format to write the heading "Is This the Best Safety Speech Ever?" which has been my most successful article (and headline) to date. It's also a great one to use for any interesting statistics you might uncover. For example:

Did You Know That More Males Die at Work Than Females?

Can You List Our Top Five Workplace Hazards?

Why is Sun Protection Important in Winter?

3. **Lists and Numbers** – People love lists because the hard work is already done for them. I've written so many blog articles with lists that I half expect someone to complain. They're also much easier to write and help structure content effortlessly. The majority of my most popular blog articles are list based. Interestingly, people prefer odd numbers to even. The numbers 3 and 7 work well, as well as 10 and 100 (these aren't odd numbers, but they do have an odd numeral in them). For example, here is a list of the top 3 blog articles I have ever written (yes, it includes an even number which defies the theory):

Ten Ways to Improve your Safety Communication

Six Tips to Improve your Toolbox Meetings

Five Mistakes Companies make with their Standard Operating Procedures

Try all of these out and see what works best for your audience. Have fun with it.

How to Choose the Right Images

By now you know how important visuals are (if not, read the previous chapter, **Persuasion**).

From now on, I want you to promise me that you will always include a visual in every single piece of safety communication you write.

Here are a few ways to incorporate photos:

- **Always have a camera available** – use your phone or a dedicated old- fashioned camera to take shots around the workplace. Where possible, include faces (preferably smiling), as we are very drawn to looking at other people. Avoid taking a photo of the back of someone. People love seeing their workmates, so if you do have a newsletter, ensure it features your staff a lot. However, be wary that if a staff member leaves, it will make those photos less powerful.

- **Use free images** – There are plenty of websites that you can go to for free photos. Just make sure you check any commercial arrangements. Some photos will require an attribution, so verify whether you need to add the photographer's name or if you have to email the photographer to ask permission. Ideally, you want to find photos that are free for commercial use.

  Free photo sites include (no license needed, but some need payment after first download):

  http://www.morguefile.com/

  http://www.freedigitalphotos.net/

http://www.stockfreeimages.com/

http://www.freemediagoo.com/

http://www.freephotosbank.com/

http://pixelperfectdigital.com/

http://openphoto.net/

http://www.stockphotosforfree.com/

- **Buy photos** – the most popular include http://www.istockphoto.com, http://www.bigstockphoto.com and http://www.gettyimages.com.

- **Use a professional photographer** – seems old-fashioned, but an experienced photographer will give you well framed shots that match your content. There's a reason why they take a few years to study the art of photography. Use a photographer if you've got a lot of important shots and you want them to last long term (for example: use in induction manuals, posters, newsletters, etc.).

- **Use ready-made icons** – again, you can go free, paid or custom made. Options include:

http://www.iconfinder.com

http://www.vecteezy.com/

http://icomoon.io/

Once you've got your photo or icon, you might need to edit them so they're the right size or even colour. Easy to use free programs include:

http://evernote.com/skitch/ (use this to annotate your photos, very handy)

http://www.sumopaint.com/

Microsoft PowerPoint can also be quite powerful to edit photos. Put your image into PowerPoint and choose the nifty little Picture Styles templates to edit what border you want around your photo. This is the program I use.

How to Structure the Main Body

Most people are time-poor, so they will quickly scan a document to see if they want to read it. The headline, image and lead will draw them in. But to keep their attention, there are some other techniques to include in your writing to hold interest and ensure an easy to scan page. These include:

- **Use a paragraph per thought** – When people see a big clump of text, they tune out and stop reading. Make your writing friendly to look at, by including lots of short paragraphs.

- **Include bullet points** – The eyes like bullet points and are drawn to them. They also make it easier for people to understand information.

- **Include a visual** – Ensure you have a photo or diagram that adds to the information. People are drawn to visuals and will look at them first.

- **Use white space** – Again, lots of text is confronting to a lot of people. Break up the text with paragraphs and spaces. Embrace white space. It also makes it easier for people to scan the article.

- **Incorporate sub headlines** – This breaks up the page and gives the eyes a break from lots of text. It also helps those who scan. You can use the headline formulas in this book for ideas. A rule of thumb is a sub headline every four paragraphs (if you don't need a sub headline, an image is also good).

- **Write concisely** - Use short sentences (17 words or less) and use short words (5 characters or less). Around 70-80% of your words should consist of only one syllable. Write at Grade 10 reading level or lower, to ensure everyone can understand. Use the readability statistics in Microsoft Word to check your writing level. For example, this book is at grade level 5.6 and readability is 66.7. This is a good writing level to aim for.

- **Make use of bold and italics** – To draw people's attention to important information, make sure you use **bold** and *italics* where you can. However, avoid using big clumps of text in either of these formats, as it is hard to read.

10 Key Tips to Write the Main Content

Now that you've got the structure in your mind and you're using the handy SELLSAFE communication template as a guide, it's time for a few points on how to improve your writing.

1. **Write with the mindset of "so what, who cares?"** Remember people are self-focused, so write what would interest them, not yourself. However, don't forget to also include the group when going deeper into the communication.

2. **Avoid rambling sentences**. Read out your words out loud and check for comprehension. Ensure that your thought process is unified and gets into all of the nitty-gritty further down the page. Delete unnecessary words or sentences that are redundant. For example: Material Safety Data Sheets are accessible in the area they are located (that's a real example from an induction manual).

3. **Where you can, show instead of tell.** If you're finding it hard to explain something, it's time to use some visuals.

4. **Ask questions in the text.** This is a good way to gain interest and get people to think. Remember, our brain is literally like a big computer, so if you ask it a question, it searches for resources and answers to that particular question. This is great for getting attention and helping people to remember the information.

5. **Avoid meaningless clichés and jargon.** They don't add anything and usually stop learning, rather than encourage.

6. **Add a story.** People love stories and a bit of mystery.

7. **Use the word 'you'.** We also prefer writing that appears to be directed at us. This pronoun gives your writing a human and warm feeling.

8. **Use positive language.** When writing your safety messages, it is important that positive language is used that focuses on the behaviour you want and not the behaviour you want to avoid. For example: "Stay Calm" is more effective than "Don't Panic." It also needs to communicate the issue in friendly language rather than a rule-based or blame-centric style. Unfortunately, around 62% of emotional words are negative. You will get little traction of your message if you blame workers for the current state of affairs. Negative words also puts images into people's heads of the wrong behaviour, which is what you need to avoid.

9. **Answer the why.** To get people to change, they need to know why. Always make sure you give reasons to why something needs to be done, so it's more compelling. It's a bit like "features and benefits." In sales writing, copywriters know that whenever they explain a product feature they must explain the benefit. For example: "our **temperature controlled cup holders** (feature) will ensure that you

beverage stays hot or cold during your trip (benefit)". If you are explaining a new safety process explain the benefit (or the why). This will also make your team smarter because if they know why something needs to be done, they will change the process, rather than leaving it in when it is no longer necessary. Remember, people always want to know "what's in it for me?" so give them a why. You can even do this by starting sentences with "You benefit by…"

For example:

Employee: "Why do I have to wear safety boots in summer?"

Safety professional: "***You benefit by*** not having your foot bitten by a snake when working in long grass."

Another way is to use the quite magical word **because**.

Remember, when you were a kid, a relevant end to an argument in the playground was "*Just because*"? In fact, when my kids have asked questions that were too hard for me to answer (okay, I was just exhausted from endless questions), I admit I've chickened out and retorted "*because*" and that was considered an effective response.

The thing is - adults are the same!

In the book, "Influence: The Psychology of Persuasion" by Robert Cialdini, he discussed a social psychology research study that found that one of the principles with human behaviour is that if we ask someone to do us a favour, we will be more successful if we provide a reason.

The magic word that made all the difference was - *because*. A research study found that saying "*Excuse me, I have five pages. May I use the Xerox machine because I'm in a rush*" had 94% effectiveness versus "*Excuse me, I have five pages.*

May I use the Xerox machine?" which was only effective 60% of the time. Further testing showed it was actually the word "*because*" that made all the difference. In fact, it didn't really matter how good the excuse was, as long as it came after the word "*because*." The word "*because*" triggers an automatic compliance response.

As a leader, include "because" whenever you need to ask people to do something - whether that means working safely or undertaking tasks in a new manner. For example, "*We need to aim for zero injuries* **because** *this means we have a safe workplace.*"

10. **End with a call to action**. Always make it clear what you want people to do. Let them know what steps to take. For example: Call your supervisor for more information or wear your hat outside every day.

4 Important Formatting Tips

In the book "Rapid Response Advertising," author Geoff Ayling lists important formatting rules to ensure that your communication is clear and easy to read.

After all, you don't want to create a brilliant campaign, only to ruin it with poor formatting.

To increase readability of your written campaign:

1. **Use Title Case** – capitals are hard to read and a lot of people think IT LOOKS LIKE YOU'RE SHOUTING AT THEM. Research shows that people shy away from reading materials with capitals. Make sure your headlines are in Title Case for best effect.

2. **Choose Sans Serif Font style** – Newspapers and books use a lot of 'serif' type, as people prefer reading this type of font.

'Serif' means flags and includes Times New Roman. If you look closely at the serif example font below, you will see the flags on each letter (such as "l" and I"). Interestingly, people are more likely to understand information if it's in Serif. Sans serif font means "without flags" such as Arial font (which is the second example).

1. Serif Example
2. Sans Serif Example

However, there is an exception to this rule. Serif fonts work well in print, but (yes, a big hairy but) sans serif font works better on computer screens. If you are writing for the intranet, use Arial (which is the most preferred).

3. **Work with Eye Gravity** – You are currently reading this book, left to right (hopefully!). This means that your eyes will tend to travel down the printed page from top left to bottom-right.

To make use of this eye-viewing pattern, make sure you put your headline at the top of the body copy (this is the main content that you write). Otherwise, people will look straight at the headline first (ignoring any information you've written before the headline). From the headline, eye gravity takes over and the eyes go down the page towards the right-hand side of the page.

However, if you are using an image, place the graphic above a headline. Our eyes are always drawn to pictures first, so if we like the picture, we then move to the headline and if that piques our interest, we then move on and scan the body copy for more information.

> Place the headline under the visual, as the eye move to the picture first and then moves down – Starch Research

4. **Use Strong Colour** – Make sure your headlines stand out in a dark colour and avoid light colours that are hard to read like yellow and pink.

Summary

One of the things that you've probably learned from this book is that advertising agencies create wonderful behavioural change campaigns because they have big budgets. They have the luxury to test the content again and again, employ graphic designers to make it all look good and copywriters to spend days tweaking content to get people to change.

While the majority of you don't have access to those resources, this book gives you the shortcuts and advertising agency secrets that you can use to make your safety communication a big success. Saving you lots of time, money and headaches.

More importantly, it will give you the skills to better influence and engage others through safety communication.

Over the next few pages are five templates that condense the information that you have learned in this book, into easy to use templates to help you create your safety communication masterpiece.

Ready to start communicating, safety artist?

PART VII

TEMPLATES

Chapter 7: 5 Templates to Make Safety Messages Stick

Download the Templates

This page is password protected.

Use password: SAFETY

http://www.digicast.com.au/safety-communication-templates/

Let's go into how you can use the following templates to allow you to quickly create safety communication that incorporates all of the elements required to change safety behaviour.

1. Employee or Contractor Persona

Before you start with any safety communication writing, fill out this easy PowerPoint document. This is in your downloadable template pack.

③-Eager Ed

BACKGROUND:
- Electrical engineer
- Thinks safety is important, but only if it doesn't take up too much time
- "We've heard this before!"

DEMOGRAPHICS:
- Skews male
- Age 30-45
- Family, 2-3 kids
- Regional

IDENTIFIERS:
- Happy demeanor
- Wants to spend time with family on weekend
- "We don't do it that way"
- Football fan

This will help you get clear on your audience, what they need to hear and how to best say it.

If you can, start by looking at the demographics (e.g., gender, age, occupation, education level) of staff from any employee satisfaction surveys that your company runs.

Otherwise, talk to HR and ask them what types of different demographics exist. If none of those are relevant to your company, then you probably have a pretty good feel for the types of groups, just by who you talk to in any safety meetings or training.

Frame the first page of this document and display on your desk to refer to when you are working.

You can also distribute to other company departments. This will impress other employees with your organisational abilities and attention to detail.

2. SELLSAFE Communication Plan Template

Use this in all your safety communication (both short and long communication programs).

If you have already filled out your employee or contractor persona, then you can use the **SELLSAFE Communication Plan** template first when you start work on a new communication piece (otherwise, go back and use the **employee persona** template before you begin).

Again, it has been designed to make you really clear on what you are trying to achieve and get you focused. It also includes all the handy tips that you have read in the book, to help remind you about the sort of content you need.

If you are working with other people, this is a great way to get you all on the same page.

3. Ad Campaign Templates or Frameworks

Use these if you are producing a workplace (or public) safety campaign and you want it to be shown over a period of time.

The templates will help you work out the best way to communicate your core message and then how to structure it. It's a great way to brainstorm communication concepts with your team.

4. Safety Communication Content Creation Template

Use this template to help structure your safety writing and to ensure you use all of the right elements such as headlines, an image and written content.

This is a good little reminder on how to make your safety communication more engaging.

5. Communication Schedule

This is a handy little Excel spreadsheet for you to fill out with all of the different formats you are going to communicate your message and when.

Communication Schedule		Insert your communication activities and the supporting materials you require. The suggestions below are intended as a guide. Highlight the amount of time required per activity.																							
Communication Activity	Supporting collateral required	Month 1				Month 2				Month 3				Month 4				Month 5				Month 6			
		Wk 1	Wk 2	Wk 3	Wk 4	Wk 1	Wk 2	Wk 3	Wk 4	Wk 1	Wk 2	Wk 3	Wk 4	Wk 1	Wk 2	Wk 3	Wk 4	Wk 1	Wk 2	Wk 3	Wk 4	Wk 1	Wk 2	Wk 3	Wk 4
Initial presentation	> PPT slides.																								
Visual reminders	> Posters																								
Personal copy to all at second presentation	> "Fold-up" pocket card																								
Reinforce PPE rules	> PPE > Posters > Notes																								
Reinforce risks and rules	> Stickers																								
Senior leaders to offer advice & feedback for toolbox talks	>																								
Review injury black spots	> Talk to employees																								
Focus on addressing injury black spots	> PPT slides																								
Publish workgroup success stories	> Communicate successes																								
Reinforcing letter from GM & pocket calendar sent home	> Pocket calendar																								

It will help you plan all the different elements that you are going to use to market your safety message over time and make it stick!

Remember, if you have a monthly theme, you need to spend as much time planning your communication as you did writing the content for your theme.

Employee or Contractor Persona

③-Eager Ed

BACKGROUND:
- Electrical engineer
- Thinks safety is important, but only if it doesn't take up too much time
- "We've heard this before!"

DEMOGRAPHICS:
- Skews male
- Age 30-45
- Family, 2-3 kids
- Regional

IDENTIFIERS:
- Happy demeanor
- Wants to spend time with family on weekend
- "We don't do it that way"
- Football fan

③-Eager Ed

GOALS:
- To do his work quickly, so he can go home on time
- To look after his mates at work

CHALLENGES:
- Prefers SMS, as email system unreliable
- Sees senior management as untrustworthy – need to get supervisor buy-in first
- Thinks safety is seen as an expense, wants management to care more

COMMUNICATION FORMATS:
- SMS safety messages and mail out monthly newsletter
- Monthly toolbox meeting best for safety theme

SELLSAFE Communication Plan Template

Objective: Get focused on what you're trying to achieve. For example: *To improve awareness of the emergency procedure* or *To improve lifting skills on the factory floor.*

Target Audience: Focus on your audience. Get into their head. If you have an employee or contractor persona filled out, transfer key information such as their image, how they like to be communicated to, and demographic data. What is it that they need to know?

Research: Common sense is the archrival of sticky safety messages. What's uncommon? What questions can you use to arouse curiosity? What are the gaps in knowledge? Uncover what's counterintuitive about the message. How can you break down people's guessing machines? Consider: "What questions do I want our staff members to ask?"

Look: What colours and consistent design elements will you use? What images are required? What visual content do you need to make?

Lasting: Communication Timeline. What media will you need to launch and when? How can you repeat your message in multiple places, multiple times? (Attach the Communication Schedule template as a separate document, if required).

Which SELLSAFE elements can you include:

Simple – What's the uncommon sense that most people don't realise? What's the one core message? What's the question you want people to ask? Write down your lead.

Emotional – What emotions does the communication invoke? If you use fear, remember to give clear steps to avoid the threat. What information is surprising? What information can you use to question what people know?

Story – What story can you use? Do you have a metaphor or safety slogan?

Authority – How can you bring credibility to your message? Can you use a real-life example, an authority figure or group? Do you have statistics to back up what you're saying?

Energy – What's the call to action? What do you want people to do with the information?

Ad Campaign Templates or Frameworks

In their research paper, The Fundamental Templates of Quality Ads," Jacob Goldenberg, David Mazursky and Sorin Solomon, from The Hebrew University of Jerusalem, found that when it comes to creating an effective ad, all award-winning ads use six templates. Even in complex thinking, certain patterns of creativity emerge.

Essentially, creative ads are more predictable than uncreative ones. If you want to spread your ideas to other people, then an easy shortcut is to work with these sets of templates that have allowed other advertising messages to succeed.

Just like an ad agency, to produce real results, you will need to understand the emotional drivers and blocks of your employees, contractors or members. Design a solid core message and safety meme (action-based) with appealing visuals.
But first, let's just take a look at the research.

The researchers assembled a group of 200 award winning, highly considered ads, and after studying them, realised that 89% could be grouped into six templates.

They then tried to fit their six templates into 200 other ads, which had not received awards. However, they were only able to classify 2% of them.

But wait, there's more.

The research team then assembled three test groups of ordinary people to create ads, which were then tested with consumers and also judged by two creative directors.

With the three test groups:

Group 1: received only two hours of training on how to use the templates
Group 2: received two hours of training with a creativity expert.
Group 3: received no training.

The group that learned about the templates was the clear winner. It was rated 50% more creative than the test groups, while consumers were also 55% more positive about the products and services that were promoted.

The good news is that there are three templates to help you create your safety campaign (from the six, I have selected the three templates more suitable for workplace safety communication).

To make these templates even more effective, remember that we are programmed **to want freedom from fear, pain and danger**. We want to work in safe workplaces where we can't get hurt. We also want to make sure our loved ones are protected.

By tapping into the very essence of what drives our behaviour, safety communication campaigns can be successfully based on the strong emotional drivers to:

- Live in physical comfort with no injuries or illness,
- Protect your child from crossing the road, swimming in a deep pool and riding their bike in traffic, and
- Work safely and be protected from potential accidents.

Remember to feature these extremely powerful emotional drivers, while using these templates.

Consequences Template

This type of safety communication suggests **the implications of either doing, or failing to perform, the recommendation** advocated in the ad.

It can often show a safety behaviour that is performed under unusual or exaggerated circumstances.

By showing an unrealistic situation, it actually helps to emphasize safety behaviours that are dangerous or inappropriate.

An example of this is the WorkSafe ad "Pain Game Worker", where we see a construction worker who is unable to access the scissor lift. He decides to take a shortcut by creating a makeshift ladder with a paint can and a box.

While drilling, a game show host and attendant magically appear, asking him to spin the wheel on the "Pain Game".

This is a great example of using a combination of surprise (we didn't expect a game show to appear on a construction site) and a metaphor to explain that taking a shortcut is a potential lottery. We then see him injured (consequences).

http://www.digicast.com.au/safety-ad-templates

You can use this template for most safety communication, but in particular if you have workers that:

- are constantly receiving the same injury,
- are undertaking a certain safety process incorrectly (despite the consequences),
- do not believe that a particular process is required,
- are not taking safety seriously,
- do not believe they could get injured.

To use this template, you will need to think about the process that you want to bring awareness to, and the really extreme consequence of not doing it. Make sure it is surprising.

Analogy Template

This is when a key attribute of a product or message is replaced with something different. It often includes an extreme replacement of what the viewer would expect.

An example of an extreme analogy is the picture I made of the shark on stairs:

As you can see, it's a pretty basic concept and easy to do.

Another good example is the Yarra Trams "Save the Rhino" campaign, which compares the heavy weight of a tram to a pack of lumbering rhinoceros.

http://www.digicast.com.au/safety-ad-templates

Use this template if you have workers that:

- are not taking safety seriously,
- are not really aware that there is a danger,
- are undertaking a certain safety process incorrectly (despite the consequences),
- do not believe that a particular process is required.

To use this template, you will need to find out more about why your employees don't believe there is a danger. Then, provide a concept that would be the extreme of what they believe. Flip their belief upside down. Make sure it is surprising.

The Altered Time Template

This style shows the outcome of using or not using a product, which is dramatised by a leap in time that shows the long-term consequences. An arresting example is the "Young Worker" ad that we discussed earlier in the book.

http://www.digicast.com.au/safety-ad-templates

In this ad, we go back in time from the present moment to see a young woman with a deformed eye waving at her children in the pool, getting married and then watch her in front of the mirror trying to conceal her facial burns. Towards the end we watch the dreadful workplace accident occur that severely injured her face. It ends with the powerful call to action: "If you're not sure, ask."

Use this template if you have workers that:

- are not taking safety seriously,
- are not really aware that there is a danger,
- are undertaking a certain safety process incorrectly (despite the consequences),
- do not believe that a particular process is required.

This template is the most extreme of all the templates and is highly fear-based. To use this, focus on a procedure you know is highly dangerous but staff seem blasé about the consequences. Make sure you have a clear call to action at the end on what to do, to ensure the fear-based ad remains highly effective.

As a word of caution, this style of ad is worth doing if you have the budget to employ a make-up artist that can create gory injuries.

Safety Communication Content Creation Template

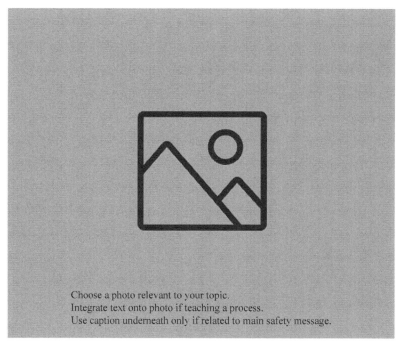

Choose a photo relevant to your topic.
Integrate text onto photo if teaching a process.
Use caption underneath only if related to main safety message.

Caption: Must include your core message

Headline (how, question, checklist)

Optional Sub-headline (how, question, checklist)

First paragraph - Start with a lead ("who, what, why, when, where" or "what, why, how") or concise clear message. Throw out everything that is meaningless and unnecessary

Second paragraph – Provide more information to detail your core message, starting with the most important. Remember, white space and one thought per paragraph. Use **bold** and *italics* to highlight important information

Use bullet points to:

- Give people clear steps
- Just break down information
- Make it easier to read

Third paragraph – Provide further detail etc. Remember, one thought per paragraph. Provide a why.

Call to action – what do you want people to do? Eg: Contact your safety advisor with any questions or to arrange training.

Communication Schedule

Communication Schedule		Insert your communication activities and the supporting materials you require. The suggestions below are intended as a guide. Highlight the amount of time required per activity.																							
		Month 1				Month 2				Month 3				Month 4				Month 5				Month 6			
Communication Activity	Supporting collateral required	Wk 1	Wk 2	Wk 3	Wk 4	Wk 1	Wk 2	Wk 3	Wk 4	Wk 1	Wk 2	Wk 3	Wk 4	Wk 1	Wk 2	Wk 3	Wk 4	Wk 1	Wk 2	Wk 3	Wk 4	Wk 1	Wk 2	Wk 3	Wk 4
Initial presentation	> PPT slides.																								
Visual reminders	> Posters																								
Personal copy to all at second presentation	> "Fold-up" pocket card																								
Reinforce PPE rules	> PPE > Posters > Notes																								
Reinforce risks and rules	> Stickers																								
Senior leaders to offer advice & feedback for toolbox talks	>																								
Review injury black spots	> Talk to employees																								
Focus on addressing injury black spots	> PPT slides																								
Publish workgroup success stories	> Communicate successes																								
Reinforcing letter from GM & pocket calendar sent home	> Pocket calendar																								

Further Reading

A Whole New Mind by Daniel H. Pink

Brain Rules by John Medina

Cashvertising by Drew Eric Whitman

Drive by Daniel H. Pink

Influence: The Psychology of Persuasion by Robert Cialdini

Inside Drucker's Brain by Jeffrey A. Krames

Made to Stick by Chip and Dan Heath

Outliers by Malcolm Gladwell

Rapid Response Advertising by Geoff Ayling

Switch by Chip and Dan Heath

The Heart of Change by John P. Kotter and Dan S. Cohen

Tested Advertising Methods by John Caples

Worldwide Rave by David Meerman Scott

"Group Performance and Decision Making," **Annual Review of Psychology, Volume 55, Norbert L. Kerr et al, 2004**

"Crafting Normative Messages to Protect the Environment," **Current Directions in Psychological Science**, 12(4), 105-109, Cialdini, R.B 2003

Let's Connect

I hope you enjoyed reading "Transform Your Safety Communication." I would love to hear from you.

Join my Facebook group page to share your journey and to connect with others about your workplace safety communication: https://www.facebook.com/groups/484522694988251/

Follow me on Twitter: https://twitter.com/digicastprodns

Connect with me on Linkedin: http://au.linkedin.com/pub/marie-claire-ross/0/260/621

Watch our videos on YouTube: https://www.youtube.com/user/Digicastprodns

If you liked this book, but want to go into deep training about each section, visit http://www.digicast.com.au/effective-communication/ to sign up for virtual group training that is held at various times throughout the year.

If you want an even easier way to keep your safety communication fresh each month, we will soon be launching a monthly publication that will give you everything you need for your safety theme. Let us know your interest at: http://www.digicast.com.au/monthly-safety-theme-communication-materials

If you want to sign up for our free newsletter (6 weekly-ish) on workplace communication, sign up at: http://www.digicast.com.au/sign-up-for-the-workplace-improver-newsletter-copy

If you want to read weekly, up to date information on safety communication and training visit our Workplace Communicator blog at: http://www.digicast.com.au/blog

If you have any questions or if you've noticed any typos or broken links, contact me directly at mc@digicast.com.au

If you liked this book, please give it a positive review on Amazon (or anywhere else!).

Thank you and I wish you all the best in you safety communication endeavours.

Yours communicatively,

Marie-Claire Ross

Free Resources

At the Digicast website (http://www.digicast.com.au) there are lots of free helpful resources that you can download. This includes:

Free Checklist: 18 Supervisor Behaviours that Produce a Thriving Culture: http://www.digicast.com.au/supervisor-checklist

8 Steps to Writing the Workplace Safety Speech: http://www.digicast.com.au/safety-speech

14 Tips to Launching a New Safety Initiative: http://www.digicast.com.au/launching-new-safety-initiative

3 Factors that Influence Workplace Culture: http://www.digicast.com.au/workplace-safety-culture

7 Communication Tips for Workplace Safety Messages
http://www.digicast.com.au/workplace-safety-messages

About the Author

 After having completed her Honors degree at Monash University in sociology (with a Psychology major), Marie-Claire began her career at top Australian market research company (ACNielsen), to research consumer and business behaviour. At AC Nielsen, she worked with well renowned advertising agencies and communication consultancies to test their communication campaigns to provide relevant advice for improved results.

Marie-Claire crafted her word-savvy skills from analyzing research data and having to communicate the results succinctly for time-poor marketing and business executives.

For 13 years, she has been an executive producer (at Digicast Productions) writing scripts for marketing and training videos in order to influence and engage, as well as articles on communication that have been published worldwide. Her popular Workplace Communicator blog is read by close to 10,000 people each month.

Marie-Claire has worked with both Australian and international companies such as CSR, Murray Goulburn, Incitec Pivot, BlueScope Steel and Country Fire Authority to create engaging videos that reduce injuries and remove the "mystery" behind procedures.

She values authentic, simple heart-felt communication and is passionate about helping safety communicators better deliver their safety messages, in order to reduce workplace risk. She's also a keen observer of human behaviour.

Marie-Claire has appeared as a guest speaker on Sky Business News and interviewed for BRW. She's also been awarded two Learnx Gold Awards for "Best Use of Video in a Training Program".

In her spare time, she loves to read business and human behaviour books, cook vegetarian food, play with her children (by getting them to read), jog and dance (but not at the same time).